Fusarium - Recent Studies

Edited by Ibrokhim Y. Abdurakhmonov

Published in London, United Kingdom

Fusarium - Recent Studies
http://dx.doi.org/10.5772/intechopen.1000249
Edited by Ibrokhim Y. Abdurakhmonov

Contributors
Abdurakhmon N. Yusupov, Abdusalom K. Makamov, Adugnaw Mekonene Admas, Albert A. Khakimov, Anvar G. Sherimbetov, Anvarjon A. Murodov, Attila Ördög, Bakhtiyor S. Adilov, Barbarita Companioni González, Batyr A. Khasanov, Bekhzod O. Mamajonov, Daniela Simões, Eugénia de Andrade, Ibrokhim Y. Abdurakhmonov, José Carlos Lorenzo Feijoo, Khurshida A. Ubaydullaeva, Malik Tahir Hayat, Mirzakamol S. Ayubov, Muhammad Nauman, Mukhtor M. Darmanov, Nadeem Iqbal, Nodirjon Y. Abdurakhmonov, Nuriddin S. Obidov, Péter Poór, Ramón Santos Bermudez, Riyazuddin Riyazuddin, Rómulo García Velasco, Shukhrat E. Shermatov, Ulmasboy T. Sobitov, Zabardast T. Buriev, Zalán Czékus, Ziyodullo H. Bashirxonov

First published in London, United Kingdom, 2024 by IntechOpen
IntechOpen is the global imprint of INTECHOPEN LIMITED, registered in England and Wales, registration number: 11086078, 167-169 Great Portland Street, London, W1W 5PF, United Kingdom

British Library Cataloguing-in-Publication Data
A catalogue record for this book is available from the British Library

Additional hard and PDF copies can be obtained from orders@intechopen.com

Fusarium - Recent Studies
Edited by Ibrokhim Y. Abdurakhmonov
p. cm.
Print ISBN 978-0-85466-488-7
Online ISBN 978-0-85466-487-0
eBook (PDF) ISBN 978-0-85466-489-4

For EU product safety concerns:
IN TECH d.o.o., Prolaz Marije Krucifikse Kozulić 3, 51000 Rijeka, Croatia,
info@intechopen.com or visit our website at intechopen.com.

Meet the editor

Ibrokhim Y. Abdurakhmonov received his BS in Biotechnology (1997) from the National University, Uzbekistan, an MS in Plant Breeding (2001) from Texas A&M University, USA, and a Ph.D. in Molecular Genetics (2002) and DSc in Genetics (2009) from the Academy of Sciences of Uzbekistan. He founded the Center of Genomics and Bioinformatics of Uzbekistan in 2012. He received the 2010 TWAS prize and ICAC Cotton Researcher of the Year 2013 for his outstanding contribution to cotton genomics and biotechnology. Dr. Abdurakhmonov was elected as a fellow of The World Academy of Sciences (TWAS) in 2014 and a member of the Academy of Sciences of Uzbekistan in 2017. In 2023, he was awarded the UNESCO-Guinea Equatorial Life Science award for contributing to molecular marker development and disease resistance in cotton.

Contents

Preface

Fusarium is one of the largest genera of filamentous fungi, represented by more than a thousand species with several continuously emerging new races and isolates. Although *Fusarium* species may cause infections in humans and animals, they are mainly soil-borne pathogens affecting plants and causing devastating crop disease and yield loss in some key crops such as cotton, barley, wheat, maize, bananas, and so on. New aggressive *Fusarium* races are appearing in the era of global climate change, prompting life scientists to develop new agro and biotechnologies to combat these new races and protect crops.

Fusarium infection in crops can be expressed in various ways, including wilting, stunting, root rot, stem rot, and discoloration of leaves and seeds. These symptoms can lead to reduced yield, poor crop quality, and economic losses for farmers. Additionally, mycotoxins produced by *Fusarium* can cause health issues if consumed by humans or animals.

Fusarium – *Recent Studies* reviews the impact of *Fusarium* pathogens in different crops. Chapters discuss new races/isolates from some crops, toxin production, plant protection mechanisms, and novel agricultural and biological control methods to combat infections. Chapters also describe identifying pathogen races in some crops and selection tools for resistant crop genotypes, which are helpful for effective crop management practices.

This book is a useful resource for agriculture university students, life science researchers, and interested readers.

I thank all chapter authors for their effort and invaluable contributions to this volume. I am thankful to the IntechOpen book department and publication managers for the opportunity to work on this book project and for their help with my editorial duties.

Ibrokhim Y. Abdurakhmonov
Center of Genomics and Bioinformatics,
Academy of Sciences of Uzbekistan,
Tashkent, Uzbekistan

Section 1

Fusiarum Diseases in Different Crops, Identification and Selection for Resistance

Chapter 1

Recent Studies on *Fusarium* Wilt in Cotton

Mirzakamol S. Ayubov, Ibrokhim Y. Abdurakhmonov, Abdusalom K. Makamov, Bekhzod O. Mamajonov, Abdurakhmon N. Yusupov, Nuriddin S. Obidov, Ziyodullo H. Bashirxonov, Anvarjon A. Murodov, Mukhtor M. Darmanov, Khurshida A. Ubaydullaeva, Shukhrat E. Shermatov, Zabardast T. Buriev, Ulmasboy T. Sobitov and Nodirjon Y. Abdurakhmonov

Abstract

Fusarium oxysporum has been a subject of intensive research since 1882, with over 200 studies published from key cotton-growing countries such as the United States, China, Uzbekistan, India, Pakistan, Australia, and Brazil. The present study has employed a diverse array of research methodologies and technological approaches, primarily emphasizing research publications disseminated within the past decade. It places specific emphasis on two key domains: Molecular Mapping and Genome-Wide Association Studies (GWAS), elucidating the evolutionary analysis transition from Simple Sequence Repeat (SSR) to Single Nucleotide Polymorphism (SNP) chip utilization. The creation of a comprehensive molecular map that incorporates Quantitative Trait Loci (QTLs) related to *Fusarium* and consolidates findings from several research groups, accompanied by figures and tables, serves to facilitate a more thorough understanding of the genetic architecture underlying *Fusarium*-related traits. An in-depth examination of recent advances in marker-assisted selection for traits conferring resistance to *Fusarium oxysporum* f. sp. *vasinfectum* (FOV), coupled with a comprehensive evaluation of the pertinent genes, offers valuable insights into the development of resistant cultivars and the underlying genetic mechanisms. This entails doing a critical review of recent relevant literature. Furthermore, this investigation examines the obstacles and potential associated with developing technologies.

Keywords: *Fusarium oxysporum*, VIGS, genes, pathogen, race

1. Introduction

Fusarium wilt is a huge challenge in cotton (*Gossypium* spp.) farming, which causes a significant reduction in yield annually. This disease, first described by Atkinson in 1882, is caused by the soil-borne pathogen *Fusarium oxysporum Schlechtend. f. sp. vasinfectum* (Atk.) [1]. Environmental parameters like as temperature, pH, light, as well as substrates all influence *Fusarium* development [2]. For example, the optimal temperature for *in vitro growth* of *F. oxysporum* isolates is between 25 and 28°C, with limits when the temperature rises above 33°C or falls below 17°C [3]. Infected plants exhibit yellowing and subsequent shedding of leaves, followed by gradual wilting and eventual demise. A distinctive indicator of fusarium wilt manifests as a reddish-brown discoloration within the vascular tissues of stems and roots, visible upon sectioning these plant parts [4].

F. oxysporum exhibits distinct variations known as "races" within its taxonomy. In the global context of cotton farming, there are six primary nominal races (different from each other by virulence rate) of FOV recognized that include 1, 2, 3, 4, 6, and 8, as documented by various researchers [5–8]. This classification was initially based on eight races but was refined through molecular characterization, revealing the merging of races 4 and 7, and races 3 and 5 [6, 9]. Geographic differences lead to distinct prevalence patterns, with race 4 causing significant losses in cotton-producing regions. In 2002, a highly virulent race 4 genotype of FOV was discovered in Californian cotton fields, marking its first appearance in the Western Hemisphere [10]. That FOV race 4 (FOV4) poses significant threats to the United States, potentially affecting the entire cotton belt, as highlighted by Zhu et al. [11].

Various studies have explored growth responses of *F. oxysporum* to metabolites of rhizospheric microflora in Egyptian cotton varieties [12]. The interaction between *Belonolaimus longicaudatus* (nematode) and FOV on cotton causes wilt development [13]. In the last few years, more cases of *Fusarium* wilt have been documented in Georgia, implying potential changes within the disease complex. During 2015 and 2016, an examination was conducted across 27 cotton fields in 10 Georgia counties. That survey involved collecting at least 10 soil and stem samples from each field from symptomatic plants to assess plant-parasitic nematode levels along with recognizing FOV races [14].

Recent years have seen a surge in scholarly articles on *Fusarium oxysporum* in cotton (*Gossypium*) species with this chapter on a comprehensive repository of relevant data on studies conducted between 2013 and 2023.

2. QTL mapping and marker-assisted selection for *Fusarium* wilt

A recent meta-analysis has examined the genetic factors associated with resistance to various stresses, both abiotic and biotic, in cotton plants. Zhang et al. in 2015 first studied a population of cotton plants known as "backcross inbred lines" (BIL) for their resistance to *Verticillium* wilt (VW) over 4 years. As a result, 10 genetic markers linked to VW resistance on cotton chromosomes, which were added to a map alongside markers, were identified. This comprehensive analysis identified 28 genetic regions, including 13 for VW, 8 for root-knot nematodes (*Meloidogyne incognita*) (RKN), and 3 for *Fusarium* wilt (FW) resistance [15].

Ulloa et al. used special cotton plant lines to investigate genetic resistance to nematodes and fungal diseases, specifically RKN and FW, respectively [16]. They found that some lines, like CS-B16, showed strong resistance to *Fusarium* race 1 due to the presence of a key genetic marker. However, other lines, such as CS-B (17-11) and CS-B17, were more susceptible to *Fusarium* race 4, indicating the presence of genes making them vulnerable. That research highlights the need for nematode-resistant cotton varieties in disease management.

In the United States, some strains of the wilt pathogen (like race 1) interact with nematodes, such as the southern RKN, causing significant root damage. When exposed to both race 1 and nematodes, certain cotton varieties exhibited moderate wilt symptoms, while another variety showed milder symptoms. Interestingly, the more virulent race 4 caused severe wilt when nematodes were present. This study revealed that race 4 can cause disease on its own and worsens when combined with nematodes, emphasizing the importance of nematode-resistant cotton varieties [17]. The variations also were explored in disease aggressiveness at different stages of seedling development. Diaz and colleagues found complex interactions between different cotton cultivars and FOV isolates, highlighting various mechanisms of resistance [18]. Additionally, a study led by Wang and colleagues [19] focused on identifying genetic markers for resistance to FOV races 1 and 4 in an interspecific cotton population. They pinpointed several key genetic markers on different chromosomes that contributed to resistance against these specific races. Studies have revealed substantial genetic diversity among FOV strains in the United States. Different genetic markers were used to categorize these strains, with the discovery of multiple genetically distinct groups. That demonstrates the diversity of FOV, particularly within the widespread race 1 genotype [20]. Furthermore, extensive trials on various cotton cultivars were conducted to assess their susceptibility to both RKN and FOV. These trials highlighted the importance of developing cotton varieties resistant to both pathogens for effective disease control [21]. Bell et al. [22] examined the genetic diversity of FOV isolates responsible for causing cotton wilt in Georgia. Researchers identified distinct vegetative complementation groups (VCGs) linked to severe outbreaks. These VCGs were associated with a specific pathotype that caused severe wilt in cotton plants. Analysis of vegetative compatibility groups (VCGs) confirmed the prevalence of one specific group (VCG01111), underscoring the absence of an exotic FOV race 4 strain in the study areas. The disease incidence increased during the 2021/22 growing season due to favorable weather conditions. Genetic analysis showed the prevalence of Australian FOV isolates, distinct from overseas races [23].

Abdelraheem and colleagues [24] used a high-density genotyping approach to study resistance to VW and virulent FOV4 in diverse cotton population groups and identified significant genetic markers correlated with resistance to VW and FOV4, laying the groundwork for potential marker-assisted breeding programs. In a greenhouse setting, FOV4-resistant and susceptible cotton plants were tested. Several genetic markers associated with resistance and demonstrated the potential for marker-assisted selection to enhance FOV4 resistance were identified [25]. Zhang et al. [26] conducted an in-depth investigation into QTLs for FOV resistance in Upland cotton, revealing 42 distinct genetic markers associated with various disease parameters. Their findings shed light on the complex genetic architecture of FOV4 resistance, emphasizing the need for multistage evaluations. In a separate study, a unique genomic region on Chromosome D03 was closely linked to stem and root vascular discoloration [27]. As found, different cotton varieties exhibited varying levels of resistance, suggesting the presence of genes responsible for resistance in this region. That work holds promise for the use of genetic markers in selecting FOV4 resistance. These studies offer valuable insights into improving cotton crop disease resistance, with implications for future breeding programs and agricultural practices [28].

3. Genes associated with cotton disease resistance

Using varying methods, scientists have undertaken a quest to identify genes related to *FOV*, an important pathogen in cotton plants. One approach is genetic and physical mapping, which has led to the discovery of resistance genes on cotton chromosomes 11 and 21 [29]. The research underscores the significance of combining genetics and genomics to pinpoint genes involved in plant resistance. To unearth genes critical to FOV infectiousness, studies have employed diverse techniques such as virus-induced gene silencing (VIGS), overexpression studies, quantitative phosphoproteomics, and bioinformatics analyses to study the Enoyl-CoA reductase (*GhECR*) gene's role in cotton's defense against different pathogenic strains [30]. In the realm of plant defense mechanisms, recognizing chitin, a key component of fungal cell walls, is crucial. In cotton, a sophisticated chitin sensing and signaling system has been identified. It involves the dimerization and phosphorylation of *GhLYK5-GhCERK1*, along with the wall-associated kinase *GhWAK7A*, which plays a significant role in plant response to infections caused by *V. dahliae* and FOV [31].

The Mitogen-Activated Protein Kinase (MAPK) cascade serves as an essential conduit for transducing external signals into internal responses to biotic and abiotic stresses. *GhMPK20*, a key gene in this cascade, has been identified as a vital player in regulating cotton resistance to *F. oxysporum*. Researchers have employed VIGS to study *GhMPK20*'s function, revealing its importance. Silencing *GhMPK20* enhanced cotton tolerance to *F. oxysporum*, while overexpression disrupted the salicylic acid (SA)-mediated defense pathway [32]. Transcriptome sequencing studies by the same researchers uncovered the essential role of group IIc *WRKY* transcription factors (TFs) in enhancing cotton resistance to FOV. That work also unveiled a novel *MAPK* cascade (*GhMKK2, GhNTF6*, and *GhMYC2*) that plays a profound role in boosting resistance by promoting the upregulation of genes involved in flavonoid biosynthesis. This, in turn, leads to flavonoid accumulation, contributing to enhanced resistance [33]. Phosphoproteomics has been used to reveal the role of *GhMORG1*, a scaffold protein, in enhancing cotton resistance to *F. oxysporum*. *GhMORG1*'s interaction with *GhMKK6* and *GhMPK4* significantly amplifies the activity of the *GhMKK6-GhMPK4* cascade. The research highlights *GhMORG1*'s involvement in coordinating various aspects of disease resistance [34]. Furthermore, Guo et al. [35] explored the *GhMKK6-GhMPK4* cascade, a critical axis in cotton resistance to FW. Silencing *GhMPK4* reduced tolerance to the disease and led to decreased expression of resistance genes. These intricate cascade attributes are akin to *GhMKK6*, adding to the complex interactions in the cotton immune response. The regulation of the *MAPK* kinase-protein phosphatase framework has been enriched with the identification of *GhAP2C1* as a negative regulator. The modulation of *GhAP2C1*'s expression revealed its dual role: silencing enhanced resistance to FW, while overexpression led to sensitivity to the disease. This highlights intricate interactions of *GhAP2C1* with *GhMPK4*, adding another layer to the complex immune response. *F. oxysporum* extracellular superoxide dismutase *FoSod5* plays a key role in pathogenicity. This protein is upregulated during infection of cotton and contributes to fungal virulence. Understanding its adaptive role in overcoming host defenses can aid in the development of effective disease management strategies [36]. These studies have illuminated the complex interplay of genes and cascades underlying defense against *F. oxysporium* in cotton, employing diverse methodologies and analytical approaches.

MicroRNAs (miRNAs) have a crucial role in regulating plant immunity and fitness. Various small RNA (sRNA) sequences, including microRNAs, during FOV pathogenesis have been identified. These sequences provide insights into plant response to the disease, with some miRNAs having roles in regulating immunity [37].

The intricate regulation of mitogen-activated protein kinase (*MAPK*) cascades in response to fungal stress remains enigmatic. Research has shed light on the collaborative role of MAPK kinase gene *GhMKK6* and microRNA ghr-miR5272a in cotton resistance to FW [38]. Plant disease resistance is a complex trait, with long noncoding RNAs (lncRNAs) playing a significant role in modulating it. Sequencing the transcriptomes of sea-island cotton recombinant inbred lines (RILs) revealed a positive correlation between the expression of FOV infection-induced lncRNAs and plant susceptibility. These lncRNAs add some complexities to plant-pathogen interactions. Long noncoding RNAs (lncRNAs) have been investigated for their role in regulating disease resistance pathways in response to *F. oxysporium* infection [39].

4. Enhancing cotton resistance to *FOV* and *V. Dahliae*

The introduction of transgenic insect-resistant cotton has raised questions about its interactions with pathogens. Gaspar and his team [40] focused on enhancing cotton resistance to *FOV* and *V. dahliae* by introducing the plant defensin *NaD1*. They created transgenic cotton plants expressing *NaD1*, which displayed robust field resistance to both pathogens. These genetically modified cotton plants exhibited increased survival rates, improved tolerance, and higher lint yield when compared to non-transgenic control counterparts. In the pursuit of boosting plant resistance against *Fusarium* wilt, Zhang et al. [24] explored innovative approaches involving the introduction of novel antimicrobial proteins like Hcm1. This approach is promising in conferring resistance to both *Verticillium* and *Fusarium* wilts in cotton by triggering innate immunity and inhibiting pathogen growth. Transgenic cotton lines expressing Hcm1 consistently exhibited higher resistance to these wilts in both greenhouse and field trials when compared to non-transgenic counterparts. Additionally, Hcm1 protein hindered the growth of *V. dahliae* and *F. oxysporum* under laboratory settings. The spread of fungal biomass was notably reduced in transgenic cotton plants as well, as the biomass of *V. dahliae* decreased significantly after inoculation with *V. dahliae* [41]. Researchers delved into the role of polygalacturonases (PGs) in the pathogenicity of *V. dahliae* and *FOV*. They cloned *VDPG1* and *FOVPG1* genes and studied their expression in different *G. hirsutum* cotton cultivars and conditions. The study revealed that *VDPG1* and *FOVPG1* were significantly elevated during infection. These findings highlight the importance of PGs in symptom development and disease progression, shedding light on the role of these enzymes in fungal pathogenicity [42]. Molecular evidence also supports the involvement of polygalacturonase-inhibiting proteins (PGIPs) in enhancing resistance to both *Verticillium* and *Fusarium* wilts. Purified *GhPGIP1*, a key defense protein, inhibits pathogenic endopolygalacturonases secreted by these fungal pathogens. Overexpressing *GhPGIP1* in transgenic Arabidopsis enhances resistance and triggers the expression of pathogenesis-related proteins (PRs) and defense-related genes [43].

Li et al. [44] investigated the physiological differences between transgenic and conventional cotton when infected by *F. oxysporum*. Their findings emphasize the impact of genetic backgrounds and modifications on the plant response to pathogens and environmental stressors. Germin-like proteins (*GLPs*) play a pivotal role in plant responses to various stresses. A novel *GLP, GhABP19*, from *G. hirsutum* has been identified and characterized. *GhABP19* is implicated in modulating resistance against both *Verticillium* and *Fusarium* wilt pathogens through its superoxide dismutase activity and regulation of jasmonic acid pathways [45]. *GLPs* are versatile glycoproteins involved in plant responses to various stresses. Another study by Pei et al. [46] emphasizes the role of *GhGLP2* in cotton's defense against *V. dahliae*, *F. oxysporum*, and oxidative stress. *GhGLP2* exhibits superoxide dismutase (*SOD*) activity and inhibits pathogen spore germination. Transgenic Arabidopsis overexpressing *GhGLP2* displays enhanced resistance to pathogens and oxidative stress, confirming its role in defense responses. Future research endeavors should focus on identifying additional factors that enhance resistance, deciphering *GLP* functions in various plant–fungal interactions, exploring ecological implications of horizontal gene transfer, and translating genomic insights into practical disease management practices. Furthermore, studies suggest that stink bugs may act as potential vectors for *FOV4* transmission. These insects can acquire and transmit the pathogen, emphasizing the importance of managing stink bugs as part of disease control strategies. That research has deepened our understanding of *Fusarium* wilt pathogenesis, including the role of specific proteins, hormonal interactions, and vector-mediated transmission. These insights provide opportunities for developing targeted strategies to manage and mitigate the impact of this devastating disease on crops [47].

5. Enhancing plant defense mechanisms

Shafique and colleagues [48] emphasize the positive impact of combining antagonistic microorganisms with organic amendments to boost plant defense mechanisms. They conducted research involving *Pseudomonas aeruginosa* and *Paecilomyces lilacinus*, combined with cotton cake, which effectively suppressed root rot fungi. This combination also stimulated the synthesis of polyphenols and antioxidants in okra plant, showcasing its potential benefits. The application of nanotechnology has enabled the creation of functionalized silver nanoparticles (AgNPs) using cellular extracts of endophytic *F. oxysporum*-NFW16. These AgNPs exhibit potent antibacterial activity against multidrug-resistant bacteria and hold promise for use in antimicrobial textile finishes [49]. Furthermore, biogenic nanoparticles, including biogenic AgNPs synthesized from extract of *F. oxysporum*, show potential for antimicrobial applications in textile and agriculture industries. Impregnating cotton fibers with these nanoparticles demonstrates their ability to prevent microbial spread, suggesting their use in antiseptic clothing and agrochemicals [50]. Cheng and colleagues [51] characterized an endochitinase (VDECH) from *V. dahliae*. Their research revealed that VDECH triggered plant defense responses and inhibited fungal spore germination, indicating its potential for disease control. The resurgence of FOV in the USA necessitates a thorough understanding of pathogenicity and host responses for effective disease prevention and management. Researchers including Cox et al. [52] explored factors contributing to disease recurrence and potential applications of existing technologies for the biological control of these pathogens. While *Fusarium* wilt is a global issue, its expansion can be managed through various approaches, including chemical and biological controls. Biological control methods have gained attention due to the adverse effects of chemical options. One study investigated the potential of the biological control agent *Saccharothrix algeriensis* NRRL B-24137 (SA) in combination with the chemical fungicide carbendazim against FOV-induced cotton wilt. The results showed significant disease control under *in vivo* conditions, highlighting the promise of integrated strategies [53]. Subsequently, the potential of the actinomycete SA as a biological control against FOV was explored. Their research included *in vitro* tests and greenhouse pot experiments, which demonstrated substantial anti-*Fusarium* activity and a reduction in disease incidence, suggesting the effectiveness of SA NRRL B-24137 in disease management [54]. Li and colleagues [55] investigated the role of salicylic acid (SA) as a biofungicide against *Fusarium* wilt. They found that SA could inhibit the pathogen by targeting the TOR (target of rapamycin) signaling pathway, offering a natural and environmentally friendly approach to disease management. Zhu et al. [56] conducted a comprehensive survey to identify and characterize FOV race 4 in New Mexico. Their research provided insights into the occurrence and virulence of FOV4 in western and southwestern regions.

The antimicrobial activities of endophytic bacteria from *Glycyrrhiza uralensis* (licorice) were examined by Mohamad et al. [57]. *Bacillus* spp., particularly *B. atrophaeus*, displayed broad-spectrum antifungal and antibacterial activity, showing promise as biocontrol agents [56]. Plant Growth-Promoting Rhizobacteria (PGPR) have gained attention for their potential to enhance plant growth and suppress fungal pathogens. Zain and colleagues [58] isolated and characterized antagonistic bacteria from cotton and sugarcane plants, highlighting their potential as biocontrol agents. *B. altitudinis* MS16 demonstrated its capability to promote mustard plant growth, while inhibiting phytopathogens like *Sclerotinia sclerotiorum* were isolated [59]. Endophytic actinobacteria associated with the medicinal plant *Thymus roseus* were found to exhibit plant growth-promoting traits, contributing to sustainable bio-fertilizer use. The antifungal activity of a cold-adapted *Pseudomonas* strain from Antarctica that produces chitinase at low temperatures was explored, potentially serving as a biocontrol agent for plant pathogens in cold environments [60]. The interactions between microbial communities in the rhizosphere and the pathobiome, as well as the role of biocontrol agents, were studied to resist pathogenic stress [61]. *Streptomyces alfalfae* was identified as a potential biocontrol agent against *Fusarium* wilt of cotton by Chen and colleagues (2021), emphasizing the multifunctionality of biocontrol agents [62]. *B. velezensis* was explored by Mohamad et al. [57] for its antimicrobial properties against viral and fungal pathogens, with potential applications in sustainable agriculture. Sahayaraj and colleagues [63] demonstrated the bioefficacy of silver nanoparticles (AgNPs) synthesized using seaweed extracts against cotton pathogenic fungi.

To overcome challenges in pathogenicity assays, an *in vitro* co-culture system was designed to study FOV infection. This foam-based system allowed controlled interaction between domesticated cotton species and FOV. The system was used to assess disease severity in different cotton cultivars, demonstrating varying responses to FOV infection. Parris and colleagues 2022, developed an *in vitro* co-culture system for studying FOV infection, providing a controlled environment for assessing disease severity in different cotton cultivars [64]. The fermentation medium for *Streptomyces alfalfae* XN-04, enhancing the production of antifungal metabolites for combatting *Fusarium* wilt in cotton, was optimized. The result showed the potential of adjusting medium components to enhance the production of antifungal metabolites that can effectively combat *Fusarium* wilt in cotton [65].

Field evaluation methods were explored for resistance to FOV race 4 in cotton breeding. The study investigated the effects of cultivar, planting date, and inoculum density on disease progression. Results revealed the impact of planting date and temperature on disease severity and the potential use of disease progression curves to assess resistance [66].

These studies collectively illustrate the potential of microbial strategies for crop protection and disease management. Utilizing antagonistic microorganisms, optimizing fermentation mediums, exploring antimicrobial metabolites, and understanding microbiota's role contribute to the advancement of sustainable agricultural practices.

6. Extracellular vesicles (EVs) in fungal systems

Extracellular vesicles (EVs) have gained considerable attention for their role in facilitating intercellular communication and macromolecule transport, especially in mammalian systems. Bleackley and colleagues [67] investigated the role of EVs in FOV infection and their involvement in the infection process. According to them, EVs isolated from the *FOV* culture medium contained unique proteins, including pigments, proteases, and polyketide synthesis-related proteins. Infiltration of FOV EVs into cotton and *N. benthamiana* leaves induced a phytotoxic response, suggesting their involvement in the infection process.

While primarily studied in human yeast pathogens, recent investigations have shed light on the involvement of fungal EVs in transporting virulence-related cargo and modulating host immune responses. Nevertheless, our understanding of EVs in filamentous fungi has been limited due to challenges such as the absence of protein markers and efficient isolation techniques. Addressing this knowledge gap, a comprehensive study has focused on the filamentous cotton pathogen FOV, providing valuable insights. The study involved isolating and characterizing the proteome of EVs derived from FOV.

To accomplish this, EVs were successfully isolated and purified by Garcia-Ceron and colleagues [68] through size-exclusion chromatography from two different growth media, Czapek Dox and Saboraud's dextrose broth. The EV proteome exhibited significant variations depending on the growth medium, while EV production remained consistent. Notably, the EVs contained proteins associated with diverse processes, including polyketide synthesis, cell wall modifications, proteases, and potential effector molecules. These findings collectively suggest the possible involvement of FOV EVs in influencing the complex interactions between the pathogen and its host, shedding light on a previously understudied aspect of host–pathogen dynamics.

In a detailed examination, the infection process of FOV4 in Pima cotton (*G. barbadense*) was dissected. Using techniques like confocal and scanning electron microscopy, the dynamics of infection by the virulent FOV4 isolate were meticulously traced in two Pima cotton varieties that included the resistant PHY 841 RF and the susceptible Pima S-7. The differences in infection were evident. PHY 841 RF displayed significantly fewer germinated conidia on its root surface, and FOV4's penetration into the root epidermis was notably delayed. Subsequent infection stages highlighted PHY 841 RF's resilience, with reduced hyphal growth and reproduction, effectively restricting the fungal ability to infiltrate the xylem. These contrasting infection dynamics provide insights into PHY 841 RF's resistance to FOV4, attributed to a series of defense mechanisms including delayed infection, inhibited fungal growth, and prevention of xylem invasion. These findings hold significance not only for comprehending cotton defense mechanisms but also for envisioning strategies to enhance crop resilience against *Fusarium* wilt [69].

7. Plant treatment in disease resistance

The application of silicon treatment has demonstrated its capacity to bolster plant resistance against various pathogens. A study delved into the influence of silicon treatment on the cellular defense responses in cotton root tissues following infection with FOV [70]. The investigation revealed that silicon treatment had a discernible impact on defense responses, especially in cotton cultivars naturally more resistant to the pathogen. Fusaric acid, a phytotoxic secondary metabolite produced by *Fusarium* spp., holds a pivotal role in pathogenicity. Another study unveiled the detoxification capabilities of the soil microbe *Mucor rouxii*, which effectively converts fusaric acid into a less harmful compound. That research highlighted the potential of microbial interactions to alleviate the adverse effects of *Fusarium* toxins on susceptible cotton plants (*G. hirsutum* and *G. barbadense*) [71].

Recent studies have explored the potential of *Talaromyces flavus* in inhibiting the growth of crucial plant pathogens, including *V. dahliae*, *Fusarium oxysporum f. sp. lycopersici*, and *Fusarium oxysporum f. sp. Cucumerinum* [72]. In response to contemporary challenges, researchers in Iran [73] have developed nano-capsules containing *T. flavus* in powder and suspension forms. Evaluations of these formulations in greenhouse conditions demonstrated their effectiveness in reducing the incidence of cotton VW.

In the context of carnation crops, *Fusarium wilt* remains a significant concern, necessitating the exploration of alternative disease management strategies. The feasibility of using near-infrared spectroscopy (NIRS) as a rapid and nondestructive method for assessing the suppressive potential of different plant growth media against carnation *Fusarium* wilt was investigated. The results indicate that NIRS holds promise for identifying suppressants, providing an efficient tool for sustainable disease management [74].

8. The complexities of *Fusarium*: host interactions and disease management strategies

Cotton cultivation faces substantial challenges due to attacks by pathogens like FOV and *V. dahliae*, which result in significant yield losses. Recent research endeavors have focused on innovative strategies for disease management. CDRAt01 is an *Aspergillus tubingensis* strain isolated from soil, known for its high tolerance to fusaric acid (FA). HPLC analysis of culture filtrates obtained from the growth of *A. tubingensis* isolate CDRAt01 in the presence of FA revealed the gradual formation of a metabolite, coinciding with a decrease in FA concentration. Subsequent spectral analysis and chemical synthesis confirmed the identity of the compound as 5-butyl-2-pyridinemethanol, commonly referred to as fusarinol. The comparative phytotoxicity assessment between fusarinol and FA was conducted by evaluating the necrotic response induced in cotyledons of cotton (*G. hirsutum* L. cv. Coker 312). Notably, fusarinol exhibited significantly lower phytotoxicity compared to FA. Consequently, the detoxification mechanism offered by the *A. tubingensis* strain presents a novel approach for mitigating *Fusarium* wilt [75]. The emergence of new FOV genotypes in China was characterized in terms of their interactions with cotton cultivars [76]. These findings underscore the necessity for ongoing surveillance to detect shifts in pathogen populations.

Research has also explored *FOV* race 3 resistances in Uzbek cotton germplasm, and genetic analyses indicated that *FOV* resistance is primarily influenced by a recessive single-gene action under high inoculum levels or disease pressure [77]. These complexities were evident in differing segregation distributions of susceptible and resistant phenotypes among Uzbek populations.

Plant–pathogen interactions are significantly shaped by phytohormones, with salicylic acid (SA) generally reducing plant susceptibility, while hormones like jasmonic acid (JA), ethylene (ET), abscisic acid (ABA), and auxins exhibit complex effects [78]. Bernardino et al. [79] showcased the antiviral activity of glucosylceramides (GlcCer) isolated from *F. oxysporum* against tobacco mosaic virus (TMV), offering a potential eco-friendly approach to enhance plant immunity. Another study by Puckhaber et al. [80] elucidated differences in active defense responses of two *G. barbadense* L. cultivars resistant to FOV race 4. The accumulation of antimicrobial terpenoids and novel compounds contributed to these resistance mechanisms [78]. Further research may explore the efficacy of AgNPs in field trials, investigate the core microbiome interactions with other pathogens, and delve deeper into noncanonical resistance genes in cotton–*FOV* interactions [81].

Understanding the molecular, genetic, and physiological aspects of plant–fungal interactions in cotton has paved the way for novel strategies to enhance disease resistance. The studies emphasize the importance of integrated approaches involving genetic studies, molecular analyses, and the utilization of beneficial microbes. These insights hold promise for the development of disease-resistant cotton cultivars, ultimately contributing to sustainable cotton production and global food security.

9. Detection and genotyping for disease management

A recent study by Zambounis and his colleagues [82] has delved into the evolutionary aspects of cotton resistance gene analogs (RGAs), unveiling signs of positive selection and potential functional implications. RGAs show great promise as functional markers in breeding programs aimed at bolstering cotton's resistance to *Fusarium* wilt. Furthermore, a specific PCR method was developed to detect FOV California race 4 by targeting a unique Tfo1 insertion event. This technique enables the timely identification of infected fields and seed lots, which is crucial for effective disease management [83]. To safeguard agricultural productivity, a comprehensive understanding of the mechanisms underlying *Fusarium*–host interactions and the development of effective disease management strategies is vital. *G. hirsutum*, susceptible to *Fusarium* wilt, often proves resilient to conventional disease management practices. Zhang et al. in 2019 delved into the population genomics of *V. dahliae*, shedding light on the defoliation phenotype and the mechanisms driving its development [84]. That study exemplifies the complexity of host–pathogen interactions and the role of horizontal gene transfer in shaping disease outcomes. Additionally, the identification and characterization of a highly virulent cotton wilt pathogen, *FOV* VCG0114 (race 4), from diverse geographical origins has provided a deep understanding on its pathogenicity and genetic diversity. The utilization of PCR-based methods for genotype determination enables effective monitoring and tracking of these devastating pathogens [85]. The sequencing of genomes of FOV isolates by Seo et al. [86] has provided an essential resource for identifying genes involved in pathogenicity and developing improved disease management strategies [84]. By using a combination of Nanopore and Illumina sequencing technology, Srivastava et al.

[87] generated genome resources for distinct FOV races, leading to concepts into their genetic makeup and unique genes. These findings hold promise for developing diagnostic markers and methods to distinguish FOV subgroups. Furthermore, *G. barbadense*, a close relative of *G. hirsutum*, faces *Fusarium* wilt challenges. Su et al. [88] uncovered the role of the anthocyanidin reductase gene (Gb_ANR-47) in enhancing resistance to *FOV* by regulating the content of proanthocyanidins. Such information highlights the intricate molecular mechanisms that contribute to plant defense responses against fungal pathogens.

To support effective disease management strategies, early detection and accurate genotyping of the pathogen are crucial. Molecular techniques, such as multiplex and singleplex PCR diagnostics, enable the identification and classification of different FOV genotypes, aiding in the prevention of further disease spread [89]. Additionally, environmental factors play a significant role in disease dynamics. DNA-based quantification and droplet digital PCR methods were used to analyze spatial variation in FOV inoculum density in soil. Their findings highlight the need to account for spatial variability in disease assessment, particularly in cultivar trials, to accurately interpret results [90].

10. The role of climate change spreading of *Fusarium* infections

Climate change profoundly impacts the distribution and severity of *Fusarium* infections. Alkhalifah et al. [91] employed geographic information system data and predictive modeling to project the future global distribution of *F. oxysporum*. The study illuminates how shifting environmental conditions under different climate scenarios could influence the prevalence of this pathogen and its associated diseases. Deletion mutants of seven genes within a single LSR (G-LSR2) were obtained from FOV using horizontal gene transfer. The findings provided compelling evidence that horizontal gene transfer from *Fusarium* to *Vd991* played an important role in its adaptation to cotton, potentially indicating a significant mechanism in the evolutionary trajectory of this asexual plant pathogen. This dynamic adaptation mechanism makes a significant contribution to *V. dahliae* virulence in cotton [92].

Lastly, susceptibility of cotton to different races of *Fusarium* wilt calls for innovative approaches to disease resistance evaluation. Zhu et al. introduced a taproot rot-based method for assessing cotton resistance to *Fusarium* wilt race 4, with findings indicating congruence between responses at the seed germination stage and subsequent stages, presenting a potential alternative for efficient and reliable resistance assessment [93]. Resequencing-based high-density genetic mapping was employed to identify QTLs for resistance to *Fusarium* wilt race 7 in *G. barbadense*, emphasizing the importance of understanding the genetic factors underlying disease resistance in cotton [94].

11. Conclusion

In summary, the comprehensive review presents a broad spectrum of research endeavors aimed at bolstering disease resistance and effectively managing the impact of pathogens, notably *FOV* in cotton. The key findings and themes encompass a range of insights. For instance, researchers have identified genetic markers associated with disease resistance in cotton, paving the way for the development of resilient cotton varieties. These markers span various chromosomes. Disease interactions can be intricate, with *RKN*s exacerbating *Fusarium* wilt symptoms, underscoring the multifaceted nature of plant–pathogen relationships. The importance of developing nematode-resistant cotton varieties is reviewed, particularly when nematodes and *FOV* jointly affect cotton plants. High-density genotyping and genetic markers have the potential to enhance disease resistance in cotton crops. Specific genomic regions closely linked to disease resistance offer the promise of selecting cotton varieties with enhanced disease resistance. Research into the molecular mechanisms of disease resistance through various techniques has provided valuable insights into the ability of cotton to fend off *FOV* and other pathogens. MAPK cascade has been identified as a key player in regulating resistance of cotton to *FOV*. Various proteins have been identified as key contributors to defense mechanisms against *FOV* and related pathogens in cotton.

Environmental factors, including climate change, significantly impact disease dynamics, affecting the distribution and prevalence of pathogens. Novel approaches for assessing resistance in cotton to diseases, such as the taproot rot-based method and high-density genetic mapping, offer efficient and reliable tools for resistance assessment. Various strategies, including silicon treatment, detoxification of phytotoxins, nanotechnology-based formulations, and early detection methods, have been explored. In conclusion, this collective body of research contributes to the ongoing mission of enhancing cotton disease resistance and mitigating the impact of devastating pathogens on cotton cultivation. These findings serve as a cornerstone for the development of sustainable agricultural practices, ultimately bolstering global food security through the cultivation of more resilient cotton varieties.

Acknowledgements

The work was done as part of the project 58-0210-9-228F "Combating *Fusarium* wilt disease, a serious economic threat for sustainable cotton production in the USA and Uzbekistan" supported by USDA-ARS. We acknowledge the Center of Genomics and Bioinformatics, Academy Sciences of the Republic of Uzbekistan research team for supporting the interpretation of the research data used in this manuscript.

Author contributions

MA, IY – coordinated, wrote, and revised the manuscript; AM, BM, AY, NO, ZB, AM, MD, KU, SS, ZB, US, and NA – collected and analyzed literature and drafted the manuscript; IA – critically read and edited the manuscript.

Ethical issues

Not applicable.

DOI: http://dx.doi.org/10.5772/intechopen.1004901

Author details

Mirzakamol S. Ayubov[1*], Ibrokhim Y. Abdurakhmonov[1], Abdusalom K. Makamov[1], Bekhzod O. Mamajonov[1], Abdurakhmon N. Yusupov[1], Nuriddin S. Obidov[1], Ziyodullo H. Bashirxonov[1], Anvarjon A. Murodov[1], Mukhtor M. Darmanov[1], Khurshida A. Ubaydullaeva[1], Shukhrat E. Shermatov[1], Zabardast T. Buriev[1], Ulmasboy T. Sobitov[2] and Nodirjon Y. Abdurakhmonov[2]

1 Center of Genomics and Bioinformatics, Academy Sciences of Uzbekistan, Tashkent, Uzbekistan

2 Institute of Soil Science and Agrochemical Research, Ministry of Agriculture of Uzbekistan, Tashkent, Uzbekistan

*Address all correspondence to: mirzo.ayubov@gmail.com

References

[1] Mokhtari S, Chavez M, Ali A. First report of *Fusarium oxysporum f. sp. vasinfectum* causing *Fusarium* wilt of cotton in Kansas, U.S.A. Plant Disease. 2023. DOI: 10.1094/PDIS-08-22-1808-PDN

[2] Mohsen AA, Abeer MS, Abeer MA. Impact of pH, temperature, light and carbon and nitrogen sources on the growth of *Fusarium oxysporum F. sp. Phaseoli*. Journal of Plant Pathology & Microbiology. 2015;**6**(7):1-7

[3] Nelson PE. History of *Fusarium* systematics. Phytopathology. 1991;**81**:1045-1048

[4] Available from: www.business.qld.gov.au/industries/farms-fishingforestry/agriculture/biosecurity/plants/diseases/horticultural/fusarium-wilt-yellows

[5] Armstrong CL, Armstrong JK. The inheritance of resistance to race 3 of *Fusarium oxysporum* in upland cotton *G. Hirsutum* L. Phytopathology. 1958;**48**(5):283-288

[6] Nirenberg HI, Feiler U, Hagedorn G. Identification of races of *Fusarium oxysporum f. sp. vasinfectum*. Phytopathology. 1994;**84**(6):653-660

[7] Kim HJ, Ji HC, Hwang SM, Yoon JB. Identification of a sequence of the formae speciales vasinfectum of *Fusarium oxysporum*. JMB. 2005;**15**(4):785-791

[8] Holmes GJ, Eckert JW, Skovgaard IM. An international study of race specificity in *Fusarium oxysporum f. sp. vasinfectum*. Phytopathology. 2009;**99**(2):239-243

[9] Skovgaard IM, Ho HY, Sackett KE, Jorgensen HJ. Genetic diversity of *Fusarium oxysporum* isolates from cotton in east and Central Africa. Phytopathology. 2001;**91**(7):684-692

[10] Puckhaber LS, Zheng X, Bell AA, Stipanovic RD, Nichols RL, Liu J, et al. Differences in active defense responses of two *Gossypium barbadense* L. cultivars resistant to *Fusarium oxysporum f. sp. vasinfectum* race 4. Journal of Agricultural and Food Chemistry. 2018a;**66**(49):12961-12966. DOI: 10.1021/acs.jafc.8b05381. Epub 2018 Dec 3

[11] Zhu Y, Abdelraheem A, Lujan P, Idowu J, Sullivan P, Nichols R, et al. Detection and characterization of *Fusarium* wilt (*Fusarium oxysporum f. sp. vasinfectum*) race 4 causing *Fusarium* wilt of cotton seedlings in New Mexico. Plant Disease. 2021;**105**(11):3353-3367. DOI: 10.1094/PDIS-10-20-2174-RE. Epub 2021 Nov 16

[12] Naim MS, Hussein AM. Growth responses of *Fusarium oxysporum* to metabolites of some rhizospheric microflora of Egyptian cotton varieties. Nature. 1958;**181**(4608):578-578. DOI: 10.1038/181578A0

[13] Yang H, Powell NT, Barker KR. Interactions of concomitant species of nematodes and *Fusarium oxysporum f. sp. vasinfectum* on cotton. Journal of Nematology. 1976;**8**:74-80

[14] da Silva MB, Davis RF, Doan HK, Nichols RL, Kemerait RC, Halpern HC, et al. *Fusarium* wilt of cotton may commonly result from the interaction of *Fusarium oxysporum f. sp. vasinfectum* with *Belonolaimus longicaudatus*. Journal of Nematology. 2019;**51**:1-10. DOI: 10.21307/jofnem-2019-015

[15] Zhang J, Yu J, Pei W, Li X, Said J, Song M, et al. Genetic analysis of *Verticillium* wilt resistance in a backcross

inbred line population and a meta-analysis of quantitative trait loci for disease resistance in cotton. BMC Genomics. 2015;**16**(1):577. DOI: 10.1186/s12864-015-1682-2

[16] Ulloa M, Wang C, Saha S, Hutmacher RB, Stelly DM, Jenkins JN, et al. Analysis of root-knot nematode and *Fusarium* wilt disease resistance in cotton (*Gossypium* spp.) using chromosome substitution lines from two alien species. Genetica. 2016;**144**(2):167-179. DOI: 10.1007/s10709-016-9887-0

[17] Wagner TA, Duke SE, Davie SM, Magill C, Liu J. Interaction of *Fusarium* wilt race 4 with root-knot nematode increases disease severity in cotton. Plant Disease. 2022;**106**(10):2558-2562. DOI: 10.1094/PDIS-12-21-2725-SC. Epub 2022 Aug 17

[18] Diaz J, Garcia J, Lara C, Hutmacher RB, Ulloa M, Nichols RL, et al. Characterization of current *Fusarium oxysporum f. sp. vasinfectum* isolates from cotton in the San Joaquin Valley of California and lower valley El Paso, Texas. Plant Disease. 2021;**105**(7):1898-1911. DOI: 10.1094/PDIS-05-20-1038-RE. Epub 2021 Aug 4

[19] Wang C, Ulloa M, Duong T, Roberts PA. Quantitative trait loci mapping of multiple independent loci for resistance to *Fusarium oxysporum f. sp. vasinfectum* races 1 and 4 in an interspecific cotton population. Phytopathology. 2018a;**108**(6):759-767. DOI: 10.1094/PHYTO-06-17-0208-R. Epub 2018 Apr 18

[20] Halpern HC, Qi P, Kemerait RC, Brewer MT. Genetic diversity and population structure of races of *Fusarium oxysporum* causing cotton wilt. G3 (Bethesda). 2020;**10**(9):3261-3269. DOI: 10.1534/g3.120.401187

[21] Wheeler TA, Dotray J, Monclova-Santana C. Effects of *Fusarium* wilt on cotton cultivars with and without Meloidogyne incognita resistance in fields. Journal of Nematology. 2022;**54**(1):20220017. DOI: 10.2478/jofnem-2022-0017

[22] Bell AA, Kemerait RC, Ortiz CS, Prom S, Quintana J, Nichols RL, et al. Genetic diversity, virulence, and Meloidogyne incognita interactions of *Fusarium oxysporum* isolates causing cotton wilt in Georgia. Plant Disease. 2017;**101**(6):948-956. DOI: 10.1094/PDIS-09-16-1382-RE. Epub 2017 Mar 14

[23] Le DP, Nguyen CPT, Kafle D, Scheikowski L, Montgomery J, Lambeth E, et al. Surveillance, diversity and vegetative compatibility groups of *Fusarium oxysporum* f. sp. vasinfectum collected in cotton fields in Australia (2017 to 2022). Pathogens. 2022;**11**(12):1537. DOI: 10.3390/pathogens11121537

[24] Abdelraheem A, Elassbli H, Zhu Y, Kuraparthy V, Hinze L, Stelly D, et al. A genome-wide association study uncovers consistent quantitative trait loci for resistance to *Verticillium* wilt and *Fusarium* wilt race 4 in the US upland cotton. Theoretical and Applied Genetics. 2020;**133**(2):563-577. DOI: 10.1007/s00122-019-03487-x. Epub 2019 Nov 25

[25] Abdelraheem A, Zhu Y, Zhang J. Quantitative trait locus mapping for *Fusarium* wilt race 4 resistance in a recombinant inbred line population of Pima cotton (*Gossypium Barbadense*). Pathogens. 2022;**11**(10):1143. DOI: 10.3390/pathogens11101143

[26] Zhang J, Abdelraheem A, Ma J, Zhu Y, Dever J, Wheeler TA, et al. Mapping of dynamic QTLs for resistance to *Fusarium* wilt (*Fusarium oxysporum*

f. sp. vasinfectum) race 4 in a backcross inbred line population of upland cotton. Molecular Genetics and Genomics. 2022a;**297**(2):319-332. DOI: 10.1007/s00438-021-01846-2. Epub 2022 Jan 12

[27] Zhang J, Zhu Y, Wheeler T, Dever JK, Hake K. Targeted development of diagnostic SNP markers for resistance to *Fusarium* wilt race 4 in upland cotton (Gossypium hirsutum). Molecular Genetics and Genomics. 2023;**298**(4):895-903. DOI: 10.1007/s00438-023-02024-2

[28] Duan X, Zhang Z, Wang J, Zuo K. Characterization of a novel cotton subtilase gene GbSBT1 in response to extracellular stimulations and its role in *Verticillium* resistance. PLoS One. 2016;**11**(4):e0153988. DOI: 10.1371/journal.pone.0153988

[29] Wang C, Ulloa M, Shi X, Yuan X, Saski C, Yu JZ, et al. Sequence composition of BAC clones and SSR markers mapped to upland cotton chromosomes 11 and 21 targeting resistance to soil-borne pathogens. Frontiers in Plant Science. 2015;**6**:791. DOI: 10.3389/fpls.2015.00791

[30] Mustafa R, Hamza M, Kamal H, Mansoor S, Scheffler J, Amin I. Tobacco rattle virus-based silencing of Enoyl-CoA reductase gene and its role in resistance against cotton wilt disease. Molecular Biotechnology. 2017;**59**(7):241-250. DOI: 10.1007/s12033-017-0014-y

[31] Wang P, Zhou L, Jamieson P, Zhang L, Zhao Z, Babilonia K, et al. The cotton wall-associated kinase *GhWAK7A* mediates responses to fungal wilt pathogens by complexing with the chitin sensory receptors. The Plant Cell. 2020a;**32**(12):3978-4001. DOI: 10.1105/tpc.19.00950

[32] Wang C, He X, Li Y, Wang L, Guo X, Guo X. The cotton MAPK kinase *GhMPK20* negatively regulates resistance to *Fusarium oxysporum* by mediating the *MKK4-MPK20-WRKY40* cascade. Molecular Plant Pathology. 2018b;**19**(7):1624-1638. DOI: 10.1111/mpp.12635

[33] Wang L, Guo D, Zhao G, Wang J, Zhang S, Wang C, et al. Group IIc WRKY transcription factors regulate cotton resistance to *Fusarium oxysporum* by promoting *GhMKK2*-mediated flavonoid biosynthesis. The New Phytologist. 2022;**236**(1):249-265. DOI: 10.1111/nph.18329. Epub 2022 Jul 27

[34] Wang C, Guo H, He X, Zhang S, Wang J, Wang L, et al. Scaffold protein *GhMORG1* enhances the resistance of cotton to *Fusarium oxysporum* by facilitating the MKK6-MPK4 cascade. Plant Biotechnology Journal. 2020b;**18**(6):1421-1433. DOI: 10.1111/pbi.13307

[35] Guo D, Hao C, Hou J, Zhao G, Shan W, Guo H, et al. The protein phosphatase *GhAP2C1* interacts together with *GhMPK4* to synergistically regulate the immune response to *Fusarium oxysporum* in cotton. International Journal of Molecular Sciences. 2022;**23**(4):2014. DOI: 10.3390/ijms23042014

[36] Wang Q, Pokhrel A, Coleman JJ. The extracellular superoxide dismutase *Sod5* from *Fusarium oxysporum* is localized in response to external stimuli and contributes to fungal pathogenicity. Frontiers in Plant Science. 2021;**12**:608861. DOI: 10.3389/fpls.2021.608861

[37] Shapulatov UM, Buriev ZT, Ulloa M, et al. Characterization of small RNAs and their targets from *Fusarium oxysporum* infected and noninfected cotton root tissues. Plant Molecular Biology Reporter. 2016;**34**(3):698-706. DOI: 10.1007/s11105-015-0945-z

[38] Wang C, He X, Wang X, Zhang S, Guo X. Ghr-miR5272a-mediated regulation of *GhMKK6* gene transcription contributes to the immune response in cotton. Journal of Experimental Botany. 2017;**68**(21-22):5895-5906. DOI: 10.1093/jxb/erx373

[39] Yao Z, Chen Q, Chen D, Zhan L, Zeng K, Gu A, et al. The susceptibility of sea-island cotton recombinant inbred lines to *Fusarium oxysporum f. sp. vasinfectum* infection is characterized by altered expression of long noncoding RNAs. Scientific Reports. 2019;**9**(1):2894. DOI: 10.1038/s41598-019-39051-2

[40] Gaspar YM, McKenna JA, McGinness BS, Hinch J, Poon S, Connelly AA, et al. Field resistance to *Fusarium oxysporum* and *Verticillium dahliae* in transgenic cotton expressing the plant defensin NaD1. Journal of Experimental Botany. 2014;**65**(6):1541-1550. DOI: 10.1093/jxb/eru021. Epub 2014 Feb 6

[41] Zhang Z, Zhao J, Ding L, Zou L, Li Y, Chen G, et al. Constitutive expression of a novel antimicrobial protein, Hcm1, confers resistance to both *Verticillium* and *Fusarium* wilts in cotton. Scientific Reports. 2016;**6**:20773. DOI: 10.1038/srep20773

[42] Liu N, Ma X, Sun Y, Hou Y, Zhang X, Li F. Necrotizing activity of *Verticillium dahliae* and *Fusarium oxysporum f. sp. vasinfectum* Endopolygalacturonases in cotton. Plant Disease. 2017a;**101**(7):1128-1138. DOI: 10.1094/PDIS-05-16-0657-RE. Epub 2017a Apr 14

[43] Liu N, Zhang X, Sun Y, Wang P, Li X, Pei Y, et al. Molecular evidence for the involvement of a polygalacturonase-inhibiting protein, *GhPGIP1*, in enhanced resistance to *Verticillium* and *Fusarium* wilts in cotton. Scientific Reports. 2017b;7:39840. DOI: 10.1038/srep39840

[44] Li X, Ding C, Wang X, Liu B. Comparison of the physiological characteristics of transgenic insect-resistant cotton and conventional lines. Scientific Reports. 2015;**5**:8739. DOI: 10.1038/srep08739

[45] Pei Y, Li X, Zhu Y, Ge X, Sun Y, Liu N, et al. *GhABP19*, a novel Germin-like protein from *Gossypium hirsutum*, plays an important role in the regulation of resistance to *Verticillium* and *Fusarium* wilt pathogens. Frontiers in Plant Science. 2019;**10**:583. DOI: 10.3389/fpls.2019.00583

[46] Pei Y, Zhu Y, Jia Y, Ge X, Li X, Li F, et al. Molecular evidence for the involvement of cotton *GhGLP2*, in enhanced resistance to *Verticillium* and *Fusarium* wilts and oxidative stress. Scientific Reports. 2020;**10**(1):12510. DOI: 10.1038/s41598-020-68943-x

[47] Esquivel JF, Bell AA. Acquisition and transmission of *Fusarium oxysporum f. sp. vasinfectum* VCG 0114 (race 4) by stink bugs. Plant Disease. 2021;**105**(10):3082-3086. DOI: 10.1094/PDIS-09-20-1999-RE

[48] Shafique HA, Sultana V, Ara J, Ehteshamul-Haque S, Athar M. Role of antagonistic microorganisms and organic amendment in stimulating the defense system of okra against root rotting fungi. Polish Journal of Microbiology. 2015;**64**(2):157-162

[49] Joshi SG, Kumar V, Janga MR, Bell AA, Rathore KS. Response of AtNPR1-expressing cotton plants to *Fusarium oxysporum f. sp. vasinfectum* isolates. Physiology and Molecular Biology of Plants. 2017;**23**(1):135-142. DOI: 10.1007/s12298-016-0411-x

[50] Ballottin D, Fulaz S, Cabrini F, Tsukamoto J, Durán N, Alves OL, et al. Antimicrobial textiles: Biogenic silver nanoparticles against *Candida* and

Xanthomonas. Materials Science & Engineering C, Materials for Biological Applications. 2017;**75**:582-589. DOI: 10.1016/j.msec.2017.02.110

[51] Cheng XX, Zhao LH, Klosterman SJ, Feng HJ, Feng ZL, Wei F, et al. The endochitinase VDECH from *Verticillium dahliae* inhibits spore germination and activates plant defense responses. Plant Science. 2017;**259**:12-23. DOI: 10.1016/j.plantsci.2017.03.002

[52] Cox KL Jr, Babilonia K, Wheeler T, He P, Shan L. Return of old foes - recurrence of bacterial blight and *Fusarium* wilt of cotton. Current Opinion in Plant Biology. 2019;**50**:95-103. DOI: 10.1016/j.pbi.2019.03.012

[53] Asif R, Siddique MH, Zakki SA, Rasool MH, Waseem M, Hayat S, et al. Saccharothrix Algeriensis NRRL B-24137 potentiates chemical fungicide Carbendazim in treating *Fusarium Oxysporum f.sp. Vasinfectum*-induced cotton wilt disease. Dose Response. 2020;**18**(3):1559325820960346. DOI: 10.1177/1559325820960346

[54] Asif R, Siddique MH, Hayat S, Rasul I, Nadeem H, Faisal M, et al. Efficacy of Saccharothrix algeriensis NRRL B-24137 to suppress *Fusarium oxysporum f.sp. vasinfectum* induced wilt disease in cotton. PeerJ. 2023;**11**:e14754. DOI: 10.7717/peerj.14754

[55] Li L, Zhu T, Song Y, Luo X, Datla R, Ren M. Target of rapamycin controls hyphal growth and pathogenicity through *FoTIP4* in *Fusarium oxysporum*. Molecular Plant Pathology. 2021;**22**(10):1239-1255. DOI: 10.1111/mpp.13108

[56] Zhu Y, Abdelraheem A, Lujan P, Idowu J, Sullivan P, Nichols R, et al. Detection and characterization of Fusarium wilt (*Fusarium oxysporum f. sp. vasinfectum*) race 4 causing *Fusarium* wilt of cotton seedlings in New Mexico. Plant Disease. 2021;**105**(11):3353-3367. DOI: 10.1094/PDIS-10-20-2174-RE. Epub 2021b Nov 16

[57] Mohamad OAA, Li L, Ma JB, Hatab S, Xu L, Guo JW, et al. Evaluation of the antimicrobial activity of endophytic bacterial populations from Chinese traditional medicinal plant Licorice and characterization of the bioactive secondary metabolites produced by bacillus atrophaeus against *Verticillium dahliae*. Frontiers in Microbiology. 2018;**9**:924. DOI: 10.3389/fmicb.2018.00924

[58] Zain M, Yasmin S, Hafeez FY. Isolation and characterization of plant growth promoting antagonistic bacteria from cotton and sugarcane plants for suppression of phytopathogenic *Fusarium* species. Iranian Journal of Biotechnology. 2019;**17**(2):e1974. DOI: 10.21859/ijb.1974

[59] Goswami M, Deka S. Isolation of a novel rhizobacteria having multiple plant growth promoting traits and antifungal activity against certain phytopathogens. Microbiological Research. 2020;**240**:126516. DOI: 10.1016/j.micres.2020.126516

[60] Liu K, Ding H, Yu Y, Chen B. A cold-adapted Chitinase-producing bacterium from Antarctica and its potential in biocontrol of plant pathogenic fungi. Marine Drugs. 2019a;**17**(12):695. DOI: 10.3390/md17120695

[61] Qiu Z, Verma JP, Liu H, Wang J, Batista BD, Kaur S, et al. Response of the plant core microbiome to *Fusarium oxysporum* infection and identification of the pathobiome. Environmental Microbiology. 2022;**24**(10):4652-4669. DOI: 10.1111/1462-2920.16194

[62] Chen J, Hu L, Chen N, Jia R, Ma Q, Wang Y. The biocontrol and plant growth-promoting properties of Streptomyces alfalfae XN-04 revealed by functional and genomic analysis. Frontiers in Microbiology. 2021;**12**:745766. DOI: 10.3389/fmicb.2021.745766

[63] Sahayaraj K, Rajesh S, Rathi JAM, Kumar V. Green preparation of seaweed-based silver nano-liquid for cotton pathogenic fungi management. IET Nanobiotechnology. 2019;**13**(2):219-225. DOI: 10.1049/iet-nbt.2018.5007

[64] Parris SM, Jeffers SN, Olvey JM, Olvey JM 2nd, Adelberg JW, Wen L, et al. An in vitro co-culture system for rapid differential response to *Fusarium oxysporum f. sp. vasinfectum* race 4 in three cotton cultivars. Plant Disease. 2022;**106**(3):990-995. DOI: 10.1094/PDIS-08-21-1743-RE

[65] Chen J, Lan X, Jia R, Hu L, Wang Y. Response surface methodology (RSM) mediated optimization of medium components for mycelial growth and metabolites production of Streptomyces alfalfae XN-04. Microorganisms. 2022;**10**(9):1854. DOI: 10.3390/microorganisms10091854

[66] Zhang J, Abdelraheem A, Zhu Y, Elkins-Arce H, Dever J, Whitelock D, et al. Studies of evaluation methods for resistance to *Fusarium* wilt race 4 (*Fusarium oxysporum f. sp. vasinfectum*) in cotton: Effects of cultivar, planting date, and inoculum density on disease progression. Frontiers in Plant Science. 2022b;**13**:900131. DOI: 10.3389/fpls.2022.900131

[67] Bleackley MR, Samuel M, Garcia-Ceron D, McKenna JA, Lowe RGT, Pathan M, et al. Extracellular vesicles from the cotton pathogen *Fusarium oxysporum f. sp. vasinfectum* induce a phytotoxic response in plants. Frontiers in Plant Science. 2020;**10**:1610. DOI: 10.3389/fpls.2019.01610

[68] Garcia-Ceron D, Dawson CS, Faou P, Bleackley MR, Anderson MA. Size-exclusion chromatography allows the isolation of EVs from the filamentous fungal plant pathogen *Fusarium oxysporum f. sp. vasinfectum* (FOV). Proteomics. 2021;**21**(13-14):e2000240. DOI: 10.1002/pmic.202000240

[69] Zhu Y, Abdelraheem A, Cooke P, Wheeler T, Dever JK, Wedegaertner T, et al. Comparative analysis of infection process in Pima cotton differing in resistance to Fusarium wilt caused by *Fusarium oxysporum f. sp. vasinfectum* race 4. Phytopathology. 2022;**112**(4):852-861. DOI: 10.1094/PHYTO-05-21-0203-R

[70] Whan JA, Dann EK, Aitken EA. Effects of silicon treatment and inoculation with *Fusarium oxysporum f. sp. vasinfectum* on cellular defences in root tissues of two cotton cultivars. Annals of Botany. 2016;**118**(2):219-226. DOI: 10.1093/aob/mcw095

[71] Crutcher FK, Puckhaber LS, Bell AA, Liu J, Duke SE, Stipanovic RD, et al. Detoxification of Fusaric acid by the soil microbe Mucor rouxii. Journal of Agricultural and Food Chemistry. 2017a;**65**(24):4989-4992. DOI: 10.1021/acs.jafc.7b01655

[72] Naraghi L, Heydari A, Rezaee S, Razavi M. Study on some antagonistic mechanisms of Talaromyces flavus against Verticillium dahliae and Verticillium alboatrum, the causal agents of wilt disease in several important crops. Biocontrol in Plant Protection. 2013;**28**(1):13-28

[73] Shabani MH, Naraghi L, Maleki M, Negahban M. Evaluation of the efficacy of different concentrations of

nano-capsules containing Talaromyces flavus with two forms of powder and suspension in reducing the incidence of cotton *Verticillium* wilt. Brazilian Journal of Biology. 2022;**84**:e262480. DOI: 10.1590/1519-6984.262480

[74] Fernández-Cabanás VM, Borrero C, Cozzolino D, Avilés M. Feasibility of near infrared spectroscopy for estimating suppressiveness of carnation (*Dianthus cariophyllus* L.) fusarium wilt in different plant growth media. Spectrochimica Acta. Part A, Molecular and Biomolecular Spectroscopy. 2022;**280**:121528. DOI: 10.1016/j.saa.2022.121528

[75] Crutcher FK, Liu J, Puckhaber LS, Stipanovic RD, Duke SE, Bell AA, et al. Conversion of fusaric acid to Fusarinol by Aspergillus tubingensis: A detoxification reaction. Journal of Chemical Ecology. 2014;**40**(1):84-89. DOI: 10.1007/s10886-013-0370-4

[76] Guo Q, Li S, Lu X, Gao H, Wang X, Ma Y, et al. Identification of a new genotype of *Fusarium oxysporum f. sp. vasinfectum* on cotton in China. Plant Disease. 2015;**99**(11):1569-1577. DOI: 10.1094/PDIS-12-14-1238-RE. Epub 2015 Sep 11

[77] Abdullaev AA, Salakhutdinov IB, Egamberdiev SS, Kuryazov Z, Glukhova LA, Adilova AT, et al. Analyses of *Fusarium* wilt race 3 resistance in upland cotton (*Gossypium hirsutum* L.). Genetica. 2015;**143**(3):385-392. DOI: 10.1007/s10709-015-9837-2

[78] Di X, Takken FL, Tintor N. How phytohormones shape interactions between plants and the soil-borne fungus *Fusarium oxysporum*. Frontiers in Plant Science. 2016;7:170. DOI: 10.3389/fpls.2016.00170

[79] Bernardino MC, Couto MLCO, Vaslin MFS, Barreto-Bergter E. Antiviral activity of glucosylceramides isolated from *Fusarium oxysporum* against tobacco mosaic virus infection. PLoS One. 2020;**15**(11):e0242887. DOI: 10.1371/journal.pone.0242887

[80] Puckhaber LS, Bell AA, Magill CW. *Fusarium* wilt race 4 (FOV4), a new race of *Fusarium oxysporum f. sp. vasinfectum* causing wilt on cotton (*Gossypium hirsutum* L.) in California. Plant Disease. 2018b;**102**(1):208-208

[81] Ilahi N, Haleem A, Iqbal S, Fatima N, Sajjad W, Sideeq A, et al. Biosynthesis of silver nanoparticles using endophytic *Fusarium oxysporum* strain *NFW16* and their in vitro antibacterial potential. Microscopy Research and Technique. 2022;**85**(4):1568-1579. DOI: 10.1002/jemt.24018

[82] Zambounis A, Ganopoulos I, Kalivas A, Tsaftaris A, Madesis P. Identification and evidence of positive selection upon resistance gene analogs in cotton (*Gossypium hirsutum* L.). Physiology and Molecular Biology of Plants. 2016;**22**(3):415-421. DOI: 10.1007/s12298-016-0362-2

[83] Ortiz CS, Bell AA, Magill CW, Liu J. Specific PCR detection of *Fusarium oxysporum f. sp. vasinfectum* California race 4 based on a unique *Tfo1* insertion event in the PHO gene. Plant Disease. 2017;**101**(1):34-44. DOI: 10.1094/PDIS-03-16-0332-RE

[84] Zhang DD, Wang J, Wang D, Kong ZQ, Zhou L, Zhang GY, et al. Population genomics demystifies the defoliation phenotype in the plant pathogen *Verticillium dahliae*. The New Phytologist. 2019;**222**(2):1012-1029. DOI: 10.1111/nph.15672

[85] Bell AA, Gu A, Olvey J, Wagner TA, Tashpulatov JJ, Prom S, et al. Detection and characterization of *Fusarium*

oxysporum f. sp. vasinfectum VCG0114 (race 4) isolates of diverse geographic origins. Plant Disease. 2019;**103**(8):1998-2009. DOI: 10.1094/PDIS-09-18-1624-RE

[86] Seo S, Pokhrel A, Coleman JJ. The genome sequence of five genotypes of *Fusarium oxysporum f. sp. vasinfectum*: A resource for studies on Fusarium wilt of cotton. Molecular Plant-Microbe Interactions. 2020;**33**(2):138-140. DOI: 10.1094/MPMI-07-19-0197-A

[87] Srivastava SK, Zeller KA, Sobieraj JH, Nakhla MK. Genome resources of four distinct pathogenic races within *Fusarium oxysporum f. sp. vasinfectum* that cause vascular wilt disease of cotton. Phytopathology. 2021;**111**(3):593-596. DOI: 10.1094/PHYTO-07-20-0298-A

[88] Su X, Zhao J, Gao W, Zu Q, Chen Q, Li C, et al. Gb_ANR-47 enhances the resistance of Gossypium barbadense to *Fusarium oxysporum f. sp. vasinfectum* (FOV) by regulating the content of Proanthocyanidins. Plants (Basel). 2022;**11**(15):1902. DOI: 10.3390/plants11151902

[89] Liu J, Wagner TA. Detection and genotyping of Fov4 (race 4, VCG0114), the *Fusarium* wilt pathogen of cotton. Methods in Molecular Biology. 2022;**2391**:191-205. DOI: 10.1007/978-1-0716-1795-3_16

[90] Davis RL 2nd, Isakeit T, Chappell TM. DNA-based quantification of *Fusarium oxysporum f. sp. vasinfectum* in environmental soils to describe spatial variation in inoculum density. Plant Disease. 2022;**106**(6):1653-1659. DOI: 10.1094/PDIS-08-21-1664-RE

[91] Alkhalifah DHM, Damra E, Melhem MB, Hozzein WN. Fungus under a changing climate: Modeling the current and future global distribution of *Fusarium oxysporum* using geographical information system data. Microorganisms. 2023;**11**(2):468. DOI: 10.3390/microorganisms11020468

[92] Chen JY, Liu C, Gui YJ, Si KW, Zhang DD, Wang J, et al. Comparative genomics reveals cotton-specific virulence factors in flexible genomic regions in *Verticillium dahliae* and evidence of horizontal gene transfer from *Fusarium*. The New Phytologist. 2018;**217**(2):756-770. DOI: 10.1111/nph.14861

[93] Zhu Y, Willey K, Wheeler T, Dever J, Whitelock D, Wedegaertner T, et al. A rapid and reliable method for evaluating cotton resistance to Fusarium wilt race 4 based on taproot rot at the seed germination stage. Phytopathology. 2023;**113**(5):904-916. DOI: 10.1094/PHYTO-08-22-0286-FI

[94] Han W, Zhao J, Deng X, Gu A, Li D, Wang Y, et al. Quantitative trait locus mapping and identification of candidate genes for resistance to *Fusarium* wilt race 7 using a resequencing-based high density genetic bin map in a recombinant inbred line population of *Gossypium barbadense*. Frontiers in Plant Science. 2022;**13**:815643. DOI: 10.3389/fpls.2022.815643

Chapter 2

Crop Diseases in Uzbekistan Caused by the Species of the Genus *Fusarium*: An Overview

Batyr A. Khasanov, Anvar G. Sherimbetov, Bakhtiyor S. Adilov and Albert A. Khakimov

Abstract

In Uzbekistan, plant diseases caused by *Fusarium* species are common. Some of them are the dominant pathogens of certain crops, causing economically important diseases such as wilts of cotton, melon, tomato and chickpea, root/crown rot of wheat, damping-off, root, crown or foot rot of pepper plants, vegetables and other crops. Reportedly, 26 *Fusarium* species cause root rot on wheat in Uzbekistan. An annotated list of 22 valid taxa is presented. As well, 26 *Fusarium* species or forms are reported on cotton. Of many *Fusaria* found on other crops, only a few credible reports have been analyzed. The composition of *Fusarium* species infecting crops still needs to be studied since their identification was based on morphology. Molecular identification started to be used recently, and so far, the results of some previous morphological identifications were confirmed, races of *Fusarium oxysporum* f. sp. *vasinfectum* and several *Fusarium* species causing root rot of wheat, bell pepper and hot pepper, chickpea, cowpea and soybean were identified. Undoubtedly, studies of *Fusarium* species, based on classical morphological methods and DNA analysis of informative genes, will be used more widely in the coming years, ensuring a reliable identification of species of this complex genus in Uzbekistan.

Keywords: wheat, cotton, vegetables, legumes, *Fusarium* diseases, species identification, morphology, DNA analysis, *TEF*-1α, ITS region, rDNA

1. Introduction

1.1 Lifestyles and importance of the *Fusarium* species

Fusarium fungi are ubiquitous worldwide and usually live in the soil as facultative plant parasites (necrotrophic pathogens) or as saprotrophs feeding on plant residues or other organic substrates. Specific populations of some *Fusarium* species, e.g., *Fusarium oxysporum* and *Fusarium solani*, can live in soil as plant pathogens or saprotrophs. Some other *Fusarium* species may exist also as endophytes of plants [1, 2].

Plant pathogenic representatives of the genus *Fusarium* can be highly aggressive and severely infect or even kill their host plants. Other *Fusarium* species may be weak (opportunistic) pathogens that infect weakened plants, aging leaves, or other parts of plants, causing only minor damage. Some species of this genus often colonize plant tissues previously affected by other pathogens (including *Fusarium* species) as secondary invaders.

Most plant pathogenic representatives of the genus *Fusarium* are widely specialized and can infect many plant species from various families. However, some species, such as *Fusarium oxysporum*, have formae speciales (f. spp.) and physiologic races, more or less narrowly specialized to infect, respectively, only the species of their host plants or their varieties [1, 2].

The negative significance of *Fusarium* fungi in people's lives is enormous and the main three reasons for this are as follows.

1. Species of this genus are pathogens of such economically important diseases as pre- and post-emergence damping-off, root rot, crown rot, foot (stem) rot, fruit rot of vegetables and other field crops, head scab of cereals.

2. The great importance of these pathogens is also evidenced by the fact that among the 10 economically most important fungi in molecular plant pathology, two species—*Fusarium graminearum* and *Fusarium oxysporum*, are placed by scientists in 4th and 5th places, respectively [3].

3. Affecting seeds of plants, especially of cereals, many *Fusarium* species can produce mycotoxins such as trichothecenes, fumonisins, zearalenones and enniatins that are dangerous to human and animal health ([4] and references therein).

4. Some *Fusarium* species are opportunistic pathogens of humans (and animals), especially with compromised immunity.

More than 70 pathogenic *Fusarium* species have been identified, which can affect people's eyes, nasal cavity, nails of hands and feet, skin, cause difficult-to-treat systemic mycoses of the lungs, intestines, bones, nervous system and refractory diseases that often lead patients to death [5–7].

1.2 Methods of identification of plant pathogenic *Fusarium* species

Due to the great importance of fungi of the genus *Fusarium*, they are objects of research by scientists in many countries. Often, the purpose of research is to determine the composition of diseases of the studied crop, for which a survey of crops, sampling of diseased plants, their mycological analysis, isolation of pure cultures of the causative agent of the disease and its identification to the level of the species is carried out.

To identify fungi, including *Fusarium* species, three main concepts used are morphological concept, biological concept and phylogenetic concept. Each concept has its advantages and disadvantages [1, 2, 8].

The morphological concept is classical, and traditional, and has been used by researchers for over two centuries [8]. It is based mainly on studying characteristics of colonies, size, shape and other characters of macroconidia, the presence or absence of microconidia and chlamydospores [1, 2]. However, it later became clear that, with rare exceptions, the morphological method does not allow differentiating most species of the genus *Fusarium* from each other [8].

The biological concept for identifying *Fusarium* species provides for crossing of the studied isolate of an unknown species with tester isolates whose species are known. Obtaining viable and fertile offspring will indicate that the studied isolate belongs to the same species as the tester isolate, crossing with which gave a positive result [1, 2]. This method is somewhat limited, since it cannot identify *Fusarium* species, in which the sexual process is unknown, nor for homothallic and some heterothallic representatives of this genus [9].

When using the molecular genetic method (in the phylogenetic concept) to identify *Fusarium* species, DNA sequences of one or more "preserved" genes in the studied isolate are determined, and with the help of the Basic Local Alignment Search Tool (BLAST) program they are compared with the available sequences in one of the databases, for example, in FUSARIUM-ID v.3 NCBI (http://isolate.fusariumdb.org/blast.php) [4–6, 10, 11].

Many "candidate" genes have been studied for molecular identification of *Fusarium* species. Although the internal transcriber spacer (ITS) region of the nuclear ribosome was selected as the universally accepted DNA barcode for fungi [12], it turned out to be less informative for the genus *Fusarium* than the other three genes, namely, partial sequences of translation elongation factor (*TEF*-1α), DNA-directed RNA polymerase II largest (*RPB* 1) and second largest (*RPB* 2) subunits [5, 13, 14]. Of these, *TEF*-1α was particularly suitable for this purpose, which alone was able to differentiate all species of the genus *Fusarium* [4, 10]. Therefore, this housekeeping gene is called the "marker of choice" for single locus identification in *Fusarium* [4, 9, 10, 13]. Based on many years of research, experienced scientists recommend using only the *TEF*-1α gene for molecular genetic identification of *Fusarium* species, and not using such genes as *ITS* rDNA (ribosomal DNA), 28S rDNA, *LSU* rDNA (large subunit ribosomal DNA) or *TUBB* (β-tubulin) [4, 10].

For identifying some other species, particularly members of the *Fusarium tricinctum* species complex, small portions of the gene encoding phosphate permease (*PHO*) enzyme were very informative [15].

Using the molecular method, more than 400 phylogenetically distinct species of the genus *Fusarium*, which belong to 23 monophyletic groups designated as "species complexes," were identified in 2022. Over 80% of these species have been discovered in the last 25 years. Due to the ongoing discovery of new species at present, this process is expected to continue in the foreseeable future [10].

Information about morphological characteristics is available for many species of the *Fusarium* genus, about biological features—for a limited number of them; there was also little information about DNA sequences of fungi of this genus; however, they are becoming more and more prevalent every year. In cases where information on all these features is available, "polyphasic" description method can be applied, which makes it possible to identify species of the genus *Fusarium* very reliably [2, 16].

Here, it is also necessary to mention the classical Koch's postulates, the fulfillment of which (i.e., the determination of the pathogenicity of the studied isolates) is always necessary in studies on the identification of plant pathogenic species of the genus *Fusarium*.

2. Diseases of crops in Uzbekistan caused by *Fusarium* species

Studies conducted in Uzbekistan revealed widespread crop diseases caused by *Fusarium* species. There are publications on wilt of cotton, root rot of wheat, vegetable crops and mulberry trees caused by species of this genus. In a number of these studies, the species of pathogens were determined tentatively—considering the species of the affected host plant and symptoms of the disease. In other studies, identification was based on the cultural and morphological characteristics of the studied isolates; in some cases, their pathogenicity to host plants was also determined. However, as mentioned above, using only the morphological concept often does not allow for the correct identification of species of this genus.

Only in recent years, to identify plant pathogenic species of the genus *Fusarium*, researchers in Uzbekistan have begun to apply methods based on the DNA analysis of certain marker genes, along with studying morphology and pathogenicity.

Below, we provide information about the *Fusarium* species, which, according to the reports of local researchers, cause diseases of crops in Uzbekistan, and discuss the status of validity of these species.

2.1 *Fusarium* species reportedly registered in Uzbekistan as wheat pathogens

As in all countries of the world, soft wheat (*Triticum aestivum* L.) in Uzbekistan is an important food crop, cultivated mainly on irrigated lands. In 2012–2021, it was sown annually on ~1.28 M hectares on average, and almost 4.68 tons of grain was harvested from each hectare [17]. Durum wheat (*Triticum durum* Desf.), previously cultivated in the country on small areas, is currently not sown.

The steady production of high grain yields in the country is hindered by wheat diseases. The most dangerous of them is stripe rust (*Puccinia striiformis* West. f. sp. *tritici* Eriks.). In some regions or in some seasons, crops may also be affected by leaf rust (*Puccinia triticina* Eriks.), common bunt (*Tilletia laevis* Kühn), loose smut (*Ustilago tritici* (Pers.) Rostr.), powdery mildew (*Blumeria graminis* (DC. ex Merat) Speer f. *tritici* É.J. Marchal), Septoria leaf blotch (*Septoria tritici* Rob. et Desm.), tan spot (*Drechslera tritici-repentis* (Died.) Shoem.) and other diseases [18–20].

In recent years, the attention of researchers has been increasingly attracted to root, crown and stem diseases of wheat plants. Fusarium root rot (FRR), Fusarium crown rot (FCR) and Fusarium head blight (FHB) are sporadically found in wheat fields in the country almost everywhere. However, in recent surveys, many fields were recorded where most of the plants have been infected at the tillering, heading and flowering growth stages. The external symptoms were stunting of plants and a general yellowing. Examination of the affected plants revealed the presence of light brown spots (necroses) on the crown and lower stem internodes. Later, the affected plants usually died prematurely, and white-head and white-stem symptoms usually appeared. In mycological analyses, fungi with morphology similar to those of *Fusarium* often grew from samples of infected plant parts. Such isolates were selected, and their species' identity was further determined, mainly by studying their morphological characteristics. In some experiments, their pathogenicity to the host plant was also determined [21, 22].

According to the literature, >60 *Fusarium* species are registered on wheat worldwide, including >40 species as causes of root and crown rot [23], and about 45 species as pathogens that infect heads [24]. According to local researchers, out of these >60 species, 26 species were registered in Uzbekistan as a cause of wheat root rot, three of which (*Fusarium avenaceum, F. graminearum* and *Fusarium poae*), in addition to the roots, also affected the ears of plants (**Tables 1** and **2**).

All species listed in **Table 1** were identified according to their morphological characteristics. Most of them were identified by Prof. A. Sheraliev [25] and his colleagues

Fusarium **species**		**Reference**
Currently accepted name	**Reported as**	
F. avenaceum	*F. avenaceum*	[21, 25]
F. buharicum	*F. bucharicum*	[25]
F. culmorum	*F. culmorum*	[21, 25–28]
F. equiseti	*F. gibbosum*	[[21, 26, 27]
F. fujikuroi	*F. fujikuroi*	[28]
F. graminearum	*F. graminearum*	[21, 25, 28]
F. heterosporum	*F. heterosporum*	[21, 25]
F. lactis	*F. lactis*	[21]
	F. moniliforme var. *lactis*	[25]
F. lateritium	*F. lateritium*	[21, 25]
F. oxysporum	*F. oxysporum*	[21, 25, 29, 30]
	F. oxysporum var. *orthoceras*	[21, 25]
F. poae	*F. poae*	[21]
	F. sporotrichiella var. *poae*	[25]
F. pseudograminearum	*F. pseudograminearum*	[29, 30]
F. redolens	*F. redolens*	[25]
F. sambucinum	*F. sambucinum*	[21, 25]
F. solani	*F. solani*	[21, 25, 29, 30]
	F. javanicum var. *radicicola*	[21, 25]
F. sporotrichioides	*F. sporotrichioides*	[28]
F. subglutinans	*F. moniliforme* var. *subglutinans*	[26, 27]
F. ventricosum	*F. solani* var. *argillaceum*	[21, 25]
F. verticillioides	*F. verticillioides*	[21]
Incertae sedis	*F. javanicum*	[21, 25]
Incertae sedis	*F. moniliforme*	[25]
Incertae sedis	*F. sambucinum* var. *ossicolum*	[25]
Not considered *Fusarium*	*F. merismoides*	[25]

Table 1.
Fusarium *species reported to infect soft and/or durum wheat plants in Uzbekistan in 1972–2019.*

Fusarium **species and strain nos.**	**Year and Region recorded**	**Isolated from**	**NCBI ID Code No.**
F. culmorum R93	2020 Tashkent	Root, crown	OP529856
F. culmorum A78	2021 Kashkadarya	Root	OP529854
F. equiseti AN5	2020 Jizzakh	Crown	OP856716
F. graminearum AN10	2020 Tashkent	Head	OP529853
F. incarnatum AN7	2021 Khorezm	Crown	OP856717
F. oxysporum AN44	2021 Tashkent	Root	OP856720

Fusarium species and strain nos.	Year and Region recorded	Isolated from	NCBI ID Code No.
F. poae AN15	2021 Tashkent	Root, crown	OP893793
F. poae AN42	2021 Tashkent	Root, crown	OP856719
F. proliferatum AN20	2020 Tashkent	Head	OP856718
F. solani AN81	2020 Tashkent	Root	OP856721
F. tricinctum AN18	2021 Andijan	Root, head	OP893792
F. tricinctum S83	2021 Fergana	Root, head	OP529855

Table 2.
Fusarium *species reported to infect soft wheat plants in Uzbekistan in 2020–2021 (unpublished data).*

[21] at the Tashkent State Agrarian University (TSAU) on the basis of morphological features of isolated fungi; for their identification, they used keys compiled by Bilai [31].

The species listed in **Table 2** were identified by the researchers of the Institute of Genetics and Plant Experimental Biology of the Academy of Sciences of the Republic of Uzbekistan (IGPEB ASU) by analysis of morphological characteristics [1], pathogenicity (**Figure 1**) and partial sequences of TEF-1α gene (unpublished data).

Twenty-two species names listed in **Tables 1** and **2** are considered valid and are part of the following species complexes (SCs): six species belong to *Fusarium sambucinum* SC (FSAMSC), five species—to *Fusarium fujikuroi* SC (FFSC), two species each to *Fusarium incarnatum-equiseti* SC (FIESC), *Fusarium solani* SC (FSSC) and *F. tricinctum* SC (FTSC), and one each to *Fusarium buharicum* SC (FBSC), *Fusarium heterosporum* SC (FHSC), *Fusarium lateritium* SC (FLSC), *Fusarium oxysporum* SC (FOSC) and to *Fusarium redolens* SC (FRSC) [10].

Figure 1.
The yellowed; dying soft wheat plants (Krasnodar 99 variety) in three pots, which soils were inoculated with the Fusarium graminearum strain AN10 (the "sickpot" trial). On the left is shown a pot with uninoculated soil and healthy plants (check treatment) (courtesy of AG Sherimbetov).

The taxonomic status of some other binomials is doubtful; these are *Fusarium javanicum* Koord. [32], *F. sambucinum* var. *ossicolum* (does not exist in Internet sources); or rejected (*Fusarium moniliforme* Sheldon nomen rejiciendum) [33]. Another species, *Fusarium merismoides* Corda, is currently transferred from the genus *Fusarium* to genus *Fusicolla* under the name of *Fusicolla merismoides* (Corda) Gräfenhan, Seifert & Schroers [34].

Below, we provide additional information about the species of the genus *Fusarium* registered in Uzbekistan on wheat. The abbreviated names of the species complexes that these species belong to are given after their names in square brackets.

Fusarium avenaceum (Fries) Sacc. [FTSC]. Many plant pathologists consider this species to be a saprotroph or weak (opportunistic) pathogen of cereals [1, 35, 36], while others [37, 38] place it among quite strong pathogens that cause wheat FRR. The high pathogenicity of *F. avenaceum* to the wheat root system is shown in experiments with artificial infection [39, 40]. Others reported that this species did not cause wheat infection in experiments [41].

In Uzbekistan, *F. avenaceum* is reported to occur sporadically in six regions of the country; its isolates from infected roots of soft and durum wheat [21, 25], as well as barley, corn, rice, etc., in total from 14 plant species, have been identified by morphological characteristics [25].

Fusarium buharicum Jacz. ex Babajan & Teteterevn.-Babajan [FBSC]. In Uzbekistan, this species was isolated from the affected roots of cotton [31] and wheat [25], and identified by morphological features. However, the pathogenicity of *F. buharicum* to wheat has not been experimentally proven, so there is no reason yet to include it among the pathogens of wheat FRR.

Fusarium culmorum (W.G. Smith) Sacc. [FSAMSC] is a strong pathogen causing FRR and FHB in soft and durum wheat, as well as cultivated and wild species of the genera *Triticum* and *Hordeum* (e.g., [1, 39, 42–50]). This species is also a pathogen of garlic, sugar beet and carnation [51].

In Uzbekistan, *F. culmorum* isolates recovered from samples infected with root rot samples of soft wheat, corn and rice collected in three regions of the country, as well as from the soil of rainfed fields, were identified by morphological features; furthermore, pathogenicity of the isolates to wheat has been confirmed experimentally [25–27].

Later, in 2020 and 2021, *F. culmorum* was isolated from the roots of soft wheat in the Tashkent and Kashkadarya regions, and two representative isolates were identified by morphological and molecular characteristics (**Table 2**). In experiments with artificial infection, these isolates showed pathogenicity to the roots of wheat plants (unpublished).

Fusarium equiseti (Corda) Sacc. [FIESC]. This species is cosmopolitan, and in mycological analysis it is very often isolated from various organs of many plants [1, 36], including infected wheat roots (e.g., [37, 42]). Recognized experts on the genus *Fusarium* (e.g., [1, 36, 42]) believe that *F. equiseti* is a common saprotroph that colonizes aging or already affected by other pathogen tissues of various plants as a secondary invader. However, in the experiments of other researchers in the USA, *F. equiseti* showed pathogenicity to seedlings of two wheat varieties [40]. It is also suggested that *F. equiseti* may be a minor pathogen of wheat FHB [43, 44].

In Uzbekistan in the 1970s, some fungi were isolated from wheat and rhizosphere soil, which were identified by morphological features as *Fusarium gibbosum* Appel & Wollenw. [26, 27]. This name is known to be synonymous with *F. equiseti*. Then, in the 2000s, *F. gibbosum* was isolated from Upland cotton, corn, etc., in total from 12 plant species in seven regions of the country [25]. However, whether all these isolates are pathogens of wheat (and other crops) has not been proven.

Later, in 2020, *F. equiseti* was isolated from the crowns of soft wheat in the Jizzakh region and was identified by morphological and molecular features (**Table 2**). In experiments with artificial infection, one isolate of the fungus showed some pathogenicity to the roots of wheat plants (unpublished).

Fusarium fujikuroi Nirenberg [FFSC]. This species, which causes the "foolish seedling" (bakanae) disease of rice, is not considered as wheat pathogen [1]. It has been, obviously, isolated by accident from wheat with root rot in the Syrdarya region of the country and was identified by the morphology of spores and molecular features [28, 52].

Fusarium graminearum Schwabe [FSAMSC]. *F. graminearum* s. str. is a cosmopolitan species, and, together with *F. culmorum* and *Fusarium pseudograminearum*, is found everywhere in the world where wheat is cultivated, but more often in regions with a temperate or warmer climate [1, 35]. It is the causal agent of very serious and economically important wheat diseases—FHB, FRR, FCR and damping-off [3]. For example, in the USA, ≥30% of the wheat grain yield is lost annually due to these diseases (e.g., [44]).

The researchers' analyses show that the causal agent of wheat FHB in the world in ~80 to 90% of cases is *F. graminearum* (and its phylogenetic "relatives," e.g., *Fusarium asiaticum* O'Donnell, T. Aoki, Kistler & Geiser). The total "share" of all other *Fusarium* species in causing FHB is estimated to be at about 10–15% [1, 43, 44, 48].

In addition to wheat species, *F. graminearum* affects other crops, and corn is the main alternative host. In susceptible varieties of this culture, the fungus strongly affects stalks and cobs (e.g., [1, 35]).

Recently, legumes have also been included among the alternative hosts of *F. graminearum, F. culmorum* and some other *Fusarium* species. Thus, isolates of these species received in the USA from chickpea and lentil infected with FRR in experiments with artificial inoculation, infected, in addition to their original host plants, also wheat and barley [53].

The registration of *F. graminearum* in eight regions of Uzbekistan on 32 crop species, including the roots and heads of soft and durum wheat, was reported back in the 2000s; identification was based on the morphology and pathogenicity of isolated strains [25]. Later, this species was also isolated by other researchers from wheat infected with FHB and FCR [21, 28].

In 2020, isolates of *F. graminearum* recovered from FHB-infected wheat heads were identified for the first time in the country on the basis of molecular genetic studies (**Table 2**; unpublished).

Fusarium heterosporum Nees ex Fries [FHSC]. Cosmopolitan, especially common in Australia and Africa on millet panicles affected by ergot [36]. It is believed that this species is not a wheat pathogen [1]. Therefore, its isolation from infected with FRR wheat in mycological analyses in Uzbekistan and elsewhere ([25, 42], a. o.) is probably associated with the invasion of the fungus into tissues as a saprotroph and secondary invader. In addition to soft and durum wheat, *F. heterosporum* was also reported in nine regions of the country on 32 crop species, including Upland and Pima-type cotton, and corn [25].

Fusarium incarnatum (Desm.) Sacc. [FIESC]. This species was isolated from a sample of the infected wheat crowns collected in the Khorezm region in 2021. A representative isolate was identified by morphological and molecular characteristics (**Table 2**).

There is no information on the Internet that *Fusarium incarnatum* can infect the roots or stems of wheat. It was only reported that this fungus is often isolated from barley and oat seeds in the north-west of Russia [54], and from wheat heads in Kenya [55]. It is believed that the ability of *F. incarnatum* to infect the roots or heads of wheat has not been experimentally proven, so it should not yet be included among the pathogens of this crop.

Fusarium lactis Pirotta & Riboni [FFSC]. This species, identified morphologically, is reported to cause FRR on wheat [21, 25] but it is a well-studied pathogen of fig trees, and it causes a serious fig disease called "endosepsis" in California, USA [1, 36]. Considering the limited area (only the USA) and the absence of pathogenicity in *F. lactis* to wheat, cotton and other plant species, one can doubt its correct identification; however, we have no reason to consider this species as a pathogen of wheat.

Fusarium lateritium Nees & Link [FLSC]. This cosmopolitan species is a parasite of trees (e.g., mulberry) on which it causes wilt, cankers and die-back, and it does not infect wheat or other field crops [1, 36]. Therefore, we assume that isolates of this species recovered from infected wheat roots in Uzbekistan [25], roots [56] and heads [57] elsewhere are probably random species that have colonized already previously infected plant tissues.

Fusarium oxysporum Schlecht. emend. Sn. & Hans. [FOSC]. This cosmopolitan fungus has both plant pathogenic and saprotrophic populations in the soil.

Wheat and other cereals are not infected by *F. oxysporum*, but this species, especially its saprobic strains, easily penetrates roots already infected by other pathogens as a secondary invader, and are easily isolated from their tissues during mycological analyses ([1].

Consequently, cultures of *F. oxysporum* recovered from wheat tissues infected with FRR/FCR in different countries, including Uzbekistan [21, 25, 29, 30] and elsewhere (e.g., [54]), are also not the true causal agents of these diseases. This has also been proved by researchers in special experiments. Thus, isolates of *F. oxysporum* and *F. solani* recovered from infected wheat roots and crowns in the USA did not show pathogenicity in tests with artificial inoculation of wheat seedlings [40].

Fusarium poae (Peck) Wollenw. [FSAMSC]. The ability of *F. poae* to cause FRR/FCR in wheat has been proven many times, but it is considered less virulent in comparison with *F. pseudograminearum, F. graminearum* and *F. culmorum* [1, 35, 38, 50].

In Uzbekistan, strains of this species isolated from wheat roots were morphologically identified under their own name, while those isolated from roots and heads—under the name *Fusarium sporotrichiella* var. *poae* [21, 25].

In 2021, *F. poae* cultures isolated from infected with FRR/FCR wheat tissues were identified for the first time in the country on the basis of molecular studies (**Table 2**; unpublished).

Fusarium proliferatum (Matsush.) Nirenberg ex Gerlach & Nirenberg [FFSC]. This cosmopolitan species is one of the aggressive pathogens causing corn stalk rot, and it also affects drooping sorghum, grasses and asparagus. Wheat is not affected by *F. proliferatum*, but it is found in its tissues as an endophyte [1, 36].

One isolate of *Fusarium proliferatum* recovered in 2020 from wheat infected with FHB was identified for the first time in Uzbekistan on the basis of molecular studies (**Table 2**; unpublished).

This species in Russia [54], China [38], Turkey [41, 42] and Iran [56] was isolated from wheat plants infected with FRR/FCR, and in some other countries—from wheat heads infected with FHB (e.g., [43, 58]). This was probably due to the transfer of *F. proliferatum* spores to wheat fields from corn or rice fields. There have been no reports of experimental establishment of the pathogenicity of *F. proliferatum* to wheat.

There was only one report to suggest that this species infected wheat leaves very weakly during artificial infection [59]. Therefore, it can be assumed that *F. proliferatum* is probably a secondary invader on the roots and heads of wheat.

Fusarium pseudograminearum Aoki & O'Donnell [FSAMSC]. This species was created in 1966 in Australia on the basis of one of the two populations of *F. graminearum* s. l., which more often cause FCR in wheat and barley, and do not form perithecia in cultures derived from a single spore. Another population of the fungus, which more often infects heads of wheat, barley, oats and corn stalks, and regularly forms perithecia in monosporic cultures, was actually *F. graminearum* s. str.

Subsequently, F. *pseudograminearum* spread from Australia to other continents [36, 60]. In recent years, it has been identified mainly as the causal agent of wheat FCR in many regions of different countries with dry and warm climates and often occupied first or second position on its incidence. It is not found in regions with cooler climates (e.g., [38, 39, 61]).

It was reported in Turdieva et al. [29] that *F. pseudograminearum* in Uzbekistan was isolated from wheat plants infected with root rot, but the identification was preliminary, not confirmed by molecular studies. For this reason, the identification of this species on wheat in Uzbekistan must be considered as yet unproven.

Fusarium redolens Wollenw. [FRSC]. This species occurs in temperate regions of the world as a saprotroph in the soil, on the rotting residue of various plants, and is the causal agent of root rot of asparagus, beans, peas, carnation, roses and spinach [1, 36]. Sometimes it is isolated from the seeds of cereals [31]. In studies conducted in the USA, *F. redolens* has shown pathogenicity to wheat seedlings [40].

It was reported that *F. redolens* has been registered in nine regions of Uzbekistan on 24 crop species, and has been isolated from soft wheat, rice, corn, Upland and Pima cottons infected with root rot. The identification of *F. redolens* was based on its morphological characteristics [25].

Fusarium sambucinum Fuck. s.str. [FSAMSC]. This species is widespread in regions of the world with a temperate, cool climate on various substrates, is often isolated from cereal seeds and causes dry rot of potato tubers [1, 36].

There are reports of isolation of *F. sambucinum* from diseased wheat roots in Iran [56] and Uzbekistan [21, 25]. The identification of the species by all these researchers was carried out on the basis of its morphology. *F. sambucinum* was included in the lists of wheat FHB pathogens [43], and isolated from the affected heads of this crop [57]; however, we have not encountered reports of experimental establishment of its pathogenicity to wheat.

Fusarium solani (Mart.) App. & Wollenw. emend Sn. et Hans. s. str., syn. *Neocosmospora solani* (Mart.) Lombard & Crous [FSSC]. This cosmopolitan fungus has plant pathogenic and saprotrophic populations in the soil. *F. solani* is the causal agent of dry and jelly-end rot of potatoes, a strong pathogen of many species of the families Fabaceae, Solanaceae, Cucurbitaceae, cotton and some trees, but it does not affect cereals, in particular, it is not the causal agent of wheat root rot [1, 32]. In experiments with artificial infection, *F. solani* showed no pathogenicity to wheat [40, 41].

Hence, it can be concluded that strains of *F. solani* isolated from the infected with FRR/FCR wheat tissues in different countries (e.g., [41, 42, 54, 56]), including Uzbekistan [21, 25], obviously are not the true causal agents of the primary root infection, but are secondary invaders.

Fusarium sporotrichioides Sherbakoff [FSAMSC]. This species occurs in regions of the world with a temperate or cold climate (even under snow). It is isolated from seeds of cereals and other plant species. In general, it is considered an opportunistic (weak, minor) pathogen of cereals (e.g., [1, 31, 36, 54]).

Fusarium sporotrichioides was isolated from infected root samples in studies conducted in Turkey [41] and Uzbekistan [28, 52]. It was identified by morphological features. The Turkish isolate did not cause wheat infection in the experiment [41]. One isolate (no. 404) in Uzbekistan was identified by molecular characteristics [52], but its pathogenicity had not been tested.

Fusarium subglutinans (Wollenw. & Reinking) Nelson, Toussoun & Marasas [FFSC]. This species causes corn stalk and cob rot in regions with a cold climate and in cooler regions of countries with a temperate climate [36]. In different countries, it was isolated from grasses, millet, drooping sorghum, wild rice and soybean [1], and in Iran [56] and Uzbekistan [26, 27]—from infected wheat roots.

Although some researchers consider *F. subglutinans* to be one of the pathogens of wheat FHB [43], there have been no reports of experimental establishment of its pathogenicity to wheat.

Fusarium tricinctum (Corda) Sacc. s. str. [FTSC] This species is more common in temperate regions of the world [1, 36]. Researchers considered it a weak pathogen, saprotroph or secondary invader, colonizing roots or heads of cereals, and sometimes wheat, previously affected by other pathogens [1, 41, 43, 44, 54, 58].

At the same time, there are reports that *F. tricinctum* is becoming a dangerous pathogen of cereals, causing FRR and FHB on wheat, which leads to severe yield losses, reduces the quality of grain, polluting it with dangerous mycotoxins such as enniatins, moniliformins, bovericin, T-2 toxin, etc. The ability of *F. tricinctum* to cause FHB in soft wheat and barley has been proven by artificial infection of plants with the isolates received from them [62–64].

In general, not only *F. tricinctum* became the emerging predominant pathogen of cereals, legumes and other economically important crops, but also other species of the *F. tricinctum* species complex [15]. *F. tricinctum* turned out to be a polyphagous that also affects worldwide oats, rice, oat, potatoes, corn, soybean, lily, onion, pumpkin and other crops [64].

In Uzbekistan, *F. tricinctum* was isolated in 2021 in the Andijan and Fergana regions from the infected FRR roots and FHB ears of wheat. Fungal isolates were identified by morphological and molecular characteristics (**Table 2**). The pathogenicity of isolates to wheat was proved in tests with artificial infection of the host plant (unpublished).

Fusarium ventricosum Appel & Wollenw., syn. *Rectifusarium ventricosum* (Appel & Wollenw.) Lombard & Crous [supposedly FSSC]. This little-known fungus was originally isolated from potato tubers and soil in a wheat field in Germany [65].

According to the report, this species (under the synonymic name *F. solani* var. *argillaceum*) was registered in eight regions of Uzbekistan on 21 plant species, including soft wheat, corn, rice and cotton. It caused root rot on wheat plants [21, 25]. Identification of this species was based on its morphological characteristics. No information is provided on testing the pathogenicity of recovered isolates to host plants. Therefore, there is no reason to consider this species a pathogen of wheat yet.

Fusarium verticillioides (Sacc.) Nirenberg, syn. *F. moniliforme* Sheldon [FFSC]. This fungus is found everywhere in the world where corn is cultivated. It is considered the most dangerous pathogen of corn, affecting its stalks and cobs. *F. verticillioides* is a polyphage, and its hosts also include drooping sorghum, sugarcane, rice, asparagus, etc., in total 11,000 plant species. However, it is believed that the pathogenicity of the fungus to these plants must be confirmed anew to make sure that no other species from the *Fusarium fujikuroi* SC is accepted for *F. verticillioides* [1, 36].

There are reports that *F. verticillioides* (sometimes given under the invalid synonymous name *F. moniliforme*) has been isolated from wheat tissues with root rot in Turkey [42, 56], Algeria [66] and Uzbekistan [21, 25]. In all these studies, identification was based on morphological features of isolates, and their pathogenicity to wheat was not tested. Therefore, it can be assumed that this species is not among the pathogens of wheat.

2.2 *Fusarium* species reportedly registered in Uzbekistan as cotton pathogens

Cotton has been cultivated in Uzbekistan for more than 2.5 thousand years and is important for the country's economy as the main cash crop. According to official sources, the total area under cotton in the country in 2019–2022 was approximately 1.1 to 1.3 million hectares annually, and the harvest ranged from 3.1 to 3.8 million tons of raw cotton.

In Uzbekistan, Upland cotton (*Gossypium hirsutum* L.) is mainly cultivated. Pima cotton (*Gossypium barbadense* L.) is grown only in the south of the country, in the Surkhandarya and Kashkadarya regions. In recent years, *G. barbadense* has occupied 4 to 5000 hectares, and in 2023 it reached 7 to 8000 hectares.

A number of abiotic and biotic factors threaten the sustainable production of high cotton yields. Of the latter, the most common and dangerous are such diseases of cotton as damping-off (**Figure 2**), root rot, Verticillium wilt, Fusarium wilt and bacterial blight. Seedling diseases and root rot diseases in the country have remained poorly studied until recent years, but wilt diseases have been known for many years and studied intensively.

Figure 2.
The dying seedling of the upland cotton (Bukhara 6 variety) infected with Fusarium sp., in the field of the Jondor district of the Bukhara region. Part of a healthy seedling is visible on the left (courtesy of BA Khasanov).

Fusarium species isolated from infected cotton roots and stems during mycological analyses in 1940 and 2001, and identified by morphological characteristics are listed in **Table 3**.

Thorough studies of the Verticillium and Fusarium wilt diseases of cotton in Uzbekistan (and in general, in Middle Asia) were conducted by A. I. Solovyova with colleagues at the Plant Protection Station of the All-Union Scientific Research Institute of Cotton Growing (PPS CRI), Tashkent region, in 1938–1945 [67–69]. During mycological analyses, the PPS employees in the 1930s recovered 25 isolates from cotton stems with wilt symptoms, which they identified by morphological features and assigned to the following 15 species: *Fusarium acuminatum*, *Fusarium anthophilum* (A. Braun) Wollenw., *Fusarium compactum* (Wollenw.) Raillo,

***Fusarium* species**		**Reference**
Currently accepted name	**Reported as**	
F. acuminatum	*F. scirpi* var. *acuminatum* (Ell. &. Ev.) Wollenw.	[67]
F. anthophilum	*F. anthophilum*	[67]
F. buharicum	*F. bucharicum*	[25]
F. compactum	*F. scirpi*. Var. *compactum* (Ell. &. Ev.) Wollenw.	[67]
F. culmorum	*F. culmorum*	[67]
F. equiseti	*F. equiseti*	[25, 67]
F. filiferum	*F. scirpi* var. *filiferum* (Preuss) Wollenw.	[67]
F. heterosporum	*F. heterosporum*	[25, 67]
F. lactis	*F. moniliforme* var. *lactis*	[25]
F. lateritium	*F. lateritium*	[25]
F. longipes	*F. scirpi* var. *longipes* (Wollenw. & Reinking) Wollenw.	[67]
F. oxysporum	*F. orthoceras* App. & Wollenw.	[67]
	F. oxysporum	[25]
F. oxysporum f. sp. *vasinfectum*	*F. vasinfectum* G.F. Atk.	[67]
F. sambucinum	*F. sambucinum*	[67]
F. scirpi	*F. scirpi*	[67]
F. solani	*F. martii* Appel & Wollenw.	[67]
	F. solani	[25, 67]
	F. javanicum var. *radicicola*	[25]
F. sporotrichioides	*F. sporotrichioides*	[67]
F. subglutinans	*F. moniliforme* var. *subglutinans*	[25]
F. ventricosum	*F. ventricosum*	[25, 67]
F. verticillioides	*F. moniliforme*	[25, 67]
Incertae sedis	*F. javanicum*	[25, 67]
Incertae sedis	*F. moniliforme* var. *minus* Wollenw. & Reinking	[67]
Not considered *Fusarium*	*F. nivale*	[25]

Table 3.
Fusarium *species reported to infect cotton plants in Uzbekistan in 1940 and 2001.*

F. culmorum, F. equiseti, Fusarium filiferum (Preuss) Wollenw., *F. heterosporum, F. longipes* Wollenw. & Reinking, *F. oxysporum, F. oxysporum* f. sp. *vasinfectum* (FOV), *F. sambucinum, Fusarium scirpi* Lamb. et Fautr., *F. solani, F. sporotrichioides, F. ventricosum, F. verticillioides* and *incertae sedis* taxa *F. javanicum* and *F. moniliforme* var. *minus*. But in the experiments of these researchers with artificial infection, none of these isolates, except FOV, infected cotton plants [67].

Subsequently, the widespread and severe infections of cotton wilt caused by FOV were also registered throughout the Andijan region [70].

Even later, A. Sheraliev from the TSAU [25] has reported that in surveys conducted on the territory of Uzbekistan, 13 *Fusarium* species were registered and identified on cotton by morphological characteristics (**Table 3**). Of these, four species were isolated from Upland cotton and nine species—from both Upland and Pima cotton species. Twelve of these 13 species were indicated by this author [25] also on wheat (see **Table 1**). In this work, it was claimed that six species, namely *F. oxysporum, F. moniliforme, F. lateritium, F. heterosporum, F. solani* and *F. equiseti*, were causal agents of cotton wilt. All these species, except *F. equiseti*, as well as five more species—*F. buharicum, F. javanicum* (*incertae sedis*), *F. lactis, F. subglutinans* (under the name of *F. moniliforme* var. *subglutinans*) and *F. ventricosum* were also indicated as pathogens causing FRR and/or FCR on cotton [25].

When analyzing the information given here, some questions arise about their validity. Thus, it is doubtful that as many as five *Fusarium* species can be true causal agents of the cotton wilt, and 11 *Fusarium* species can be true causal agents of FRR/FCR on cotton. Currently, only *Fusarium oxysporum* f. sp. *vasinfectum* is recognized as the only causative agent of cotton wilt [1, 71].

In addition, another species, *Fusarium nivale* Ces. ex Berl. & Voglino, was indicated as a pathogen of Upland cotton, as well as corn and five other plant species [25]. This species is excluded from the genus *Fusarium*, and its current name is *Microdochium nivale* (Fr.) Samuels & I.C. Hallett [1]. Due to the fact that *M. nivale* causes pink snow rot of cereals usually in regions with a cold climate [72], the report of its presence on cotton in hot and arid conditions of Uzbekistan does not seem convincing.

Obviously, all of the above information indicates once more that the morphological identification of species of this genus is often unreliable and can lead to confusion. Therefore, in order to obtain reliable results in the differentiation of fungi, in particular plant pathogenic *Fusarium* species, along with traditional methods of morphological identification and determination of their pathogenicity, it is necessary to use molecular genetic methods [10, 73].

After the statement by A. Sheraliev [25] that *Fusarium moniliforme* is one of the causal agents of the cotton wilt, another publication claimed that this fungus allegedly causes wilting on almost all varieties of *Gossypium hirsutum* deployed in various zones, and this disease is widespread in 10 regions of Uzbekistan; the identification of *F. moniliforme* was based on its morphological characteristics [74]. However, considering the fact that the name *Fusarium moniliforme* has long become illegitimate [33], researchers [75, 76] have questioned the reliability of this message. Later, these authors [77] started to use a name *F. verticillioides* instead of *F. moniliforme*, and its teleomorph stage they called *Gibberella fujikuroi* (Sawada) Ito. In fact, the teleomorph of *F. verticillioides* is *Gibberella moniliformis* Wineland, and not *G. fujikuroi* (Sawada) Ito in Ito & K. Kimura [1]. In that paper, the authors also claimed that they conducted molecular studies, and "genes from *NIR* and rDNA were sequenced and turned out to be similar to the genes of *Gibberella fujikuroi* (Sawada) Ito, which is the sexual stage of

Fusarium moniliforme Sheld." [77]. The details of the study of these genes, the results of comparing their sequences with those of *F. verticillioides* in this work were not presented, and we also did not find them in the National Center for Biotechnology Information, USA (NCBI) GenBank database. For this reason, it is necessary to consider the statement that *F. verticillioides* is the causative agent of cotton wilt as unproven.

Further studies with the causal agents of Fusarium wilt and root rot of cotton were carried out by employees of the Center for Genomics and Bioinformatics of the Academy of Sciences of Uzbekistan (the CGB ASU) [52, 78, 79]. In their experiments, they used more than 45 isolates from the collection of the Institute of Genetics and Plant Experimental Biology of the Academy of Sciences of Uzbekistan (IGPEB ASU), 34 of which were isolated in 2002–2013 from symptomatic cotton plants in different regions of Uzbekistan and identified by morphological characteristics according to Booth [80]. The nucleotide sequences of the gene translation elongation factor 1α (*TEF*-1α), β-tubulin (*TUBB*), nitrate reductase (*NIR*), ribosomal DNA (rDNA) and phosphate permease (*PHO*) were determined in these isolates and compared with the information of the NCBI GenBank using the BLAST method. The studies also included 13 FOV isolates with known races, which they received from the USA [52, 78].

In these studies, of the 34 isolates studied, 14 were identified as *F. oxysporum* f. sp. *vasinfectum* (FOV), 3—as *F. solani*, 1 each—as *F. oxysporum*, *F. oxysporum* f. sp. *phaseoli* and *F. verticillioides*, 5—as *F. proliferatum* and 3—as *F. fujikuroi* (**Table 4**). Isolates 319 and 552 could not be accurately identified. The results of morphological and molecular identification coincided for eight FOV isolates (nos. 30, 34, 316, 328, 347,

Isolate No. and region collected	Morphological Identification	Molecular Identification	Reference
30 Andijan	FOV	FOV	[79]
34 Kashkadarya	FOV	FOV	[79]
316 Tashkent	FOV	FOV, race 3	[52, 78]
328 Bukhara	FOV	FOV, race 7	[52]
		FOV, race 4 and 7	[78]
347 Syrdarya	FOV	FOV, race 3	[52, 79]
375 Tashkent	*F. oxysporum*	*F. oxysporum*	[52]
		FOV, race 1, 2 and 6	[78]
444 Tashkent	*F. oxysporum*	FOV	[52]
		Fusarium sp.	[78]
460, 491 Andijan	*F. oxysporum*	FOV	[52]
489 Bukhara	FOV	FOV, race 1	[52]
496 Kashkadarya	*F. oxysporum*	FOV, race 4	[52]
526 Bukhara	FOV	FOV, race 4	[52]
527 Tashkent	*Fusarium* sp.	FOV, race 3	[79]
534 Andijan	FOV	FOV	[52]
489 Bukhara	FOV / *Fusarium* sp.	*F. oxysporum*	[52]
139 Surkhandarya	*F. oxysporum*	*Fusarium equiseti*	[52]

Isolate No. and region collected	Morphological Identification	Molecular Identification	Reference
315, 327 Unknown	*F. oxysporum*	*Fusarium equiseti*	[52]
498, 525 Kashkadarya	*F. sambucinum*	*Fusarium equiseti*	[52]
528 Tashkent	*F. verticillioides*	*Fusarium equiseti*	[52]
319 Namangan	*F. oxysporum*	*F. solani*	[52, 78]
		Fusarium sp.	[78]
319.1 Namangan	*F. solani*	*F. solani*	[79]
519, 520 Bukhara	*F. solani*	*F. solani*	[52]
490 Syrdarya	*F. oxysporum* / *F. verticillioides*	*F. proliferatum*	[52]
502 Tashkent	*F. solani*	*F. proliferatum*	[52]
509 Tashkent	*F. solani*	*F. proliferatum*	[52]
	F. oxysporum	*F. oxysporum* f. sp. *phaseoli*	[52]
518 Tashkent	*F. oxysporum*	*F. proliferatum*	[52]
524 Tashkent	*F. oxysporum* / *F. verticillioides*	*F. proliferatum*	[52]
552 Tashkent	*Fusarium* sp.	*F. proliferatum* or *F. fujikuroi*	[52]
487 Tashkent	*F. verticillioides*	*F. fujikuroi*	[52]
559 Kashkadarya	*F. lateritium*	*F. fujikuroi*	[52]
561 Tashkent	*Fusarium* sp.	*F. fujikuroi*	[52]
602 Bukhara	*F. verticillioides*	*F. verticillioides*	[52]

Table 4.
Identification of Fusarium *isolates recovered from upland cotton and some other crops in Uzbekistan in 2013–2016.*

489, 526 and 534), three isolates of *F. solani* (nos. 319.1, 519 and 520) and one isolate of *F. verticillioides* (no. 602). It was also found that FOV populations in Uzbekistan are represented by complex race 1, 2 & 6, race 3 and race 4 or 7 [52, 78, 79].

It is puzzling here that at every mention of the *F. verticillioides* (isolate nos. 487, 490, 524, 528 and 602), the name *Gibberella fujikuroi* is mistakenly given as its teleomorph stage. In addition, without questioning the reliability and significance of the results obtained in these studies, it should also be noted that we could not find in the NCBI GenBank database at least some information about the species of the genus *Fusarium* identified by the authors using molecular method [52, 78, 79].

Pathogenicity of the individual representative isolates of *Fusarium* species was studied in laboratory, greenhouse and field experiments. The ability to cause wilting of cotton plants has been confirmed for a number of FOV isolates. *F. solani* caused root rot of cotton seedlings, but its separate isolates differed greatly in the degree of pathogenicity: some of them were avirulent or showed moderate pathogenicity, while others severely affected cotton plants, often killing them. *Fusarium proliferatum* and *F. equiseti* isolates did not show pathogenicity to cotton plants in tests, and the pathogenicity of *F. fujikuroi* and *F. verticillioides* isolates was not studied [52, 78].

2.3 *Fusarium* species reportedly registered in Uzbekistan as pathogens of other crops

According to the results of many years of research, A. Sheraliev [25] has reported the isolation and morphological identification using keys of Bilai [31] of 17 species and 10 varieties of *Fusarium* fungi from 57 species of agricultural crops and weeds. However, due to the fact that major changes have taken place in the taxonomy and nomenclature of the genus *Fusarium* in recent years, the status of many taxa identified by this researcher requires clarification. In addition, some of his statements raise doubts about their reliability; for example, it was stated that as many as 10 *Fusarium* species cause wilt, and as many as 14 *Fusarium* species cause root rot on mulberry trees (*Morus* spp.) [25, 81].

Studies conducted in Uzbekistan have shown that *Fusarium* species cause economically important diseases not only on cotton and wheat, but also on a number of other agricultural crops (**Table 5**). It was found that most of these diseases are caused by *F. oxysporum* or its formae speciales. Such diseases included wilts of melon and tomato, root and crown rots of pepper and some other crops. Below is a summary of these and other diseases registered in the country.

Isolate No. and region collected	Host	Morphological Identification	Molecular Identification and NCBI ID No.	Reference
– All regions	Misc. crops	Misc. species	—	[25]
	Melon	*F. oxysporum* f. sp. *melonis*	—	[82]
	Tomato	*F. oxysporum* f. sp. *lycopersici*	—	[83, 84]
	Misc. crops	*F. oxysporum*	—	[85]
	Potato, onion	*F. solani*	—	[85]
– Kashkadarya	Bell pepper	*F. solani*	—	[86]
– Tashkent	Bell pepper	*F. solani*	—	[86–88]
– Kashkadarya	Bell pepper	*F. oxysporum*	—	[86–88]
– Fergana	Chili pepper	*F. oxysporum*	—	[86]
– Tashkent	Chili pepper	*F. oxysporum*	—	[86]
– Surkhandarya	Rice	*Fusarium* sp.	—	[89]
– Tashkent	Radishes	*F. oxysporum* f. sp. *raphani*	—	[90]
28 Tashkent	Melon	*F. oxysporum*	*F. equiseti*	[52]
378 Jizzakh	Bindweed	*F. oxysporum* / *F. longipes*	*F. equiseti*	[52]
546 Tashkent	Tomato	*F. oxysporum* / *F. verticillioides*	*F. proliferatum*	[52]
547 Tashkent	Tomato	*F. oxysporum* / *F. verticillioides*	*F. proliferatum*	[52]
		F. oxysporum	*F. oxysporum* f. sp. *phaseoli*	[79]
560 Tashkent	Tomato	*Fusarium* sp.	*F. fujikuroi*	[52]

Isolate No. and region collected	Host	Morphological Identification	Molecular Identification and NCBI ID No.	Reference
UZF-21 Andijan	Bell pepper	*F. oxysporum*	*F. oxysporum* OQ466308	IGPEB*
UZF-23 Andijan	Bell pepper	*F. oxysporum*	*F. oxysporum* OQ466309	IGPEB
R-22 Karakalpakstan	Cowpea	*F. oxysporum*	*F. oxysporum* OP716769	IGPEB
R-23 Tashkent	Soybean	*F. acuminatum*	*F. acuminatum* OP716770	IGPEB
R-24 Syrdarya	Soybean	*F. acuminatum*	*F. acuminatum* OP716771	IGPEB
R-25 Jizzakh	Chickpea	*F. equiseti*	*F. equiseti* OP716774	IGPEB
R-29 Tashkent	Soybean	*F. culmorum*	*F. brachygibbosum* OP716772	IGPEB
R-37 Tashkent	Chickpea	*F. camptoceras*	*F. persicinum* OP716773	IGPEB

**IGPEB – group of scientists of the IGPEB ASU (unpublished).*

Table 5.
Identification of Fusarium *isolates recovered from some other crops in Uzbekistan.*

Melon wilt caused by *F. oxysporum* f. sp. *melonis* Snyder & Hansen is widespread almost everywhere in Uzbekistan and is the main disease of this crop. The disease is severe and causes large crop losses, especially when there is a lack of soil moisture. Often in such cases, most of the plants of susceptible melon varieties are killed in the late season, usually 15–25 days before the full ripening of the fruit [82, 83].

It was reported that tomato wilt, caused by *F. oxysporum* f. sp. *lycopersici* (Sacc.) Snyder & Hansen, in Uzbekistan is among the dominant diseases of this crop both in the open fields and in greenhouses [83, 84]. However, there are no exact data on the incidence of the disease in different regions of the country.

Recently, a frequent occurrence in the Tashkent region of wilt disease on green and red globe radishes caused by *F. oxysporum* f. sp. *raphani* Kendrick & Snyder was reported in Akbarov and Khakimova [90].

It is also indicated that in Uzbekistan *F. oxysporum* (the form is not specified) causes wilt diseases on cucumbers, pumpkins and watermelons, and rotting of carrot fruits, and *F. solani* causes rotting of potato tubers and onion heads [85].

In 2010, in the paddies of "Sholi Invest" LLC in the Surkhandarya region, spots on panicles and spots and lesions on rice seeds were encountered. On the panicles just emerging from the sheathes of the flag leaf, 5–10% to 60–80% of the seeds already had symptoms of the disease. At the request of farmers, samples of the affected plant parts were collected, which were examined visually and subjected to mycological analysis. *Fusarium* sp. was consistently isolated from ~60% and *Curvularia lunata* (Wakk.) Boed. from ~15% of the samples. The symptoms of the disease on the panicles and seeds were typical of the disease called "pecky rice" [91]. It is caused by various fungi in different countries of the world, among which *C. lunata* [92] and several species of the genus *Fusarium* [93] are often found.

In recent years, cases of sudden withering and death of sweet and hot pepper plants (*Capsicum annuum* L.) have been registered in six districts of four regions of Uzbekistan. The number of dead plants in these areas ranged from 1 to 3 to 5% in the greenhouses, and from 5 to 21.1% in the open fields [86–88]. Examinations showed the presence of two groups of symptoms on plants. The signs of the disease (**Figure 3**) on the plants of the first group were very similar to those on the crown rot of pepper,

Figure 3.
The stem rot of hot pepper plant (Shakira F_1 variety), caused by Fusarium oxysporum in a field of the Kibray district of Tashkent region: A general view of the affected, withered, and healthy plants (at center), and their lower parts, and roots, and the cross-sections through the areas of their crowns (courtesy of AA Khakimov).

found earlier and described as a new pathogen in Spain, named *F. oxysporum* f. sp. *radicis-capsici* Lomas-Cano et al. [94, 95].

This form of the disease was found on seedlings and young plants in all surveyed regions of Uzbekistan. From the symptomatic crown samples of plants collected in Uzbekistan, colonies of the same morphology were constantly isolated, representative isolates from which were identified as *F. oxysporum* [86–88].

On plants of the second group, the disease was represented by stem rot, which led to sudden withering and rapid death of abundantly fruiting plants (**Figure 4**). This disease was recorded in one field in Kashkadarya, and one in Tashkent regions, where the number of dead plants averaged 11.9 and 10.5%, respectively [87, 88]. The symptoms of the disease were similar to those of the stem rot of sweet pepper caused by *F. solani* in Canada [96], but we did not see perithecia of the pathogen on stems in Uzbekistan. The pathogen from symptomatic plants was isolated and tentatively identified by morphological features as *F. solani* [87, 88].

Currently, marker gene sequences are being analyzed with Uzbek isolates of *F. oxysporum* and *F. solani* from pepper plants to confirm the results of morphological identification [86–88].

Reportedly, in recent years, several *Fusarium* species (**Table 5**) have been isolated and identified from samples of diseased plants collected in Uzbekistan, by morphological and molecular characteristics, including *F. equiseti* from melon and bindweed, *F. proliferatum* and *F. fujikuroi* from tomato [52] and *F. oxysporum* f. sp. *phaseoli* Kendrick & Snyder from tomato plants [79]. Experiments to confirm the pathogenicity of these species to host plants have not been conducted.

Another group of researchers at the Institute of Genetics and Plant Experimental Biology of the Academy of Sciences of the Republic of Uzbekistan (IGPEB ASU)

Figure 4.
The stem rot of hot pepper plants (Lastochka variety), a general view of the affected and withered plants in the field of the Yakkabag district of Kashkadarya region; arrows show infection sites (courtesy of BA Khasanov).

isolated and identified eight strains belonging to five species of the genus *Fusarium* from samples of root rot-infected crops collected in 2022 in five regions of Uzbekistan by morphological and molecular characters (analysis of DNA sequences of ITS region rDNA) (**Table 5**) (unpublished). *F. oxysporum* was isolated from samples of dead or dying seedlings taken from two greenhouses in the Andijan region producing sweet pepper seedlings (**Figure 5**). The same species was isolated from symptomatic young cowpea plants, samples of which were collected in Karakalpakstan. Isolates of *F. acuminatum* were isolated from samples of soybean plants affected by root rot in the Tashkent and Syrdarya regions, and another species, *Fusarium brachygibbosum* Padwick [FSAMSC], was also isolated from Tashkent samples. The latter is a polyphage and an emerging new root/crown pathogen of wheat, legumes, corn and other crops and trees; it has been registered in Algeria, Iraq and some other countries of the world ([97], and references therein). From the samples of symptomatic chickpea plants collected in the Jizzakh and Tashkent regions, *F. equiseti* and *Fusarium persicinum* J. W. Xia et al. (FIESC) were isolated, respectively.

Fusarium acuminatum and *F. equiseti* in North America and elsewhere are included in the complex of fungi that cause root rot of legumes [53, 98]. The pathogenicity of Uzbek isolates of *F. oxysporum* to pepper and cowpea plants, *F. equiseti* to chickpeas, and *F. acuminatum* and *F. brachygibbosum* to soybean plants has been confirmed in experiments with artificial infection. There is no information about the pathogenicity of *F. persicinum* to plants. It is known to be a mycoparasite on other fungi, more often on basidiomycetes [99].

Figure 5.
A focus of the seedling damping-off of the sweet pepper plants (Ferrari F_1-Sakata variety), caused by Fusarium oxysporum and Rhizoctonia solani complex in the greenhouse's nursery of the Shakhrikhon district of the Andijan region (courtesy of DT Aznabakieva).

3. Conclusions

Diseases of crops caused by fungi from the genus *Fusarium* are common in Uzbekistan and are of great economic importance. Lists of species of this genus have been compiled that were reportedly registered in Uzbekistan on soft and/or durum wheat (26 species, including 22 valid names), Upland and/or Pima-type cotton (26 species or varieties, including 21 proper names) and other crops (> 11 species or varieties). Based on a critical analysis of these reports, the following conclusions can be drawn. The species recorded on wheat in Uzbekistan include well-known and highly aggressive pathogens *Fusarium graminearum* and *F. culmorum*, an emerging pathogen *F. tricinctum*, and less potent pathogens *F. avenaceum* and *F. poae*. The finding in Uzbekistan of *F. pseudograminearum*, the strong wheat crown rot pathogen, is considered not yet confirmed. Possibly, some strains of *F. sporotrichioides*, *F. equiseti* and *Fusarium redolens* could be weak pathogens of wheat, but confirmation of their pathogenicity in experiments with artificial infection of the host plant is required. Several other *Fusarium* species, namely *F. buharicum, F. fujikuroi, F. incarnatum, F. lactis, F. oxysporum, F. sambucinum, F. solani, F. subglutinans, F. verticillioides*, and, probably, *F. ventricosum*, reportedly isolated from wheat plants in the country, are obviously not wheat pathogens, but are secondary invaders.

Among the *Fusarium* species isolated by various researchers from cotton in Uzbekistan, the pathogenicity to the host plant has been confirmed for many strains of FOV (cause of wilt), and several isolates of *F. solani* (cause of damping-off and FRR). By studying the nucleotide sequences of the *TEF*-1α, *TUBB* and rDNA genes, it was found that the FOV populations in Uzbekistan are represented by complex race 1, 2 & 6, race 3 and race 4 or 7.

Reports that the causal agents of cotton wilt are, in addition to *F. oxysporum*, five more species, namely, *F. equiseti, F. heterosporum, F. lateritium, F. solani* and *F. verticillioides* (reported as *F. moniliforme* nom. Invalid.), have no justification. So, in the experiments with artificial infection, cotton isolates of *F. equiseti* and *F. proliferatum* did not show pathogenicity to this culture. The report that *Fusarium verticillioides* (the sexual stage of which is erroneously listed as *Gibberella fujikuroi*) causes cotton wilt in many regions of Uzbekistan is also not confirmed by convincing results of molecular studies. Also unproven are the claims that the same five species mentioned above, as well as *F. buharicum*, *F. javanicum* (*incertae sedis*), *F. lactis* and *F. ventricosum*, are among the pathogens of cotton root rot. Besides, the indication of *Microdochium nivale* (reported as *Fusarium nivale*) as a pathogen of cotton and other crops is also not credible.

In addition to wheat and cotton, *Fusarium* spp. in Uzbekistan cause serious diseases on other crops. Surveys conducted in recent years have shown that such diseases caused by *F. oxysporum* and other *Fusarium* species include wilt of melons and other cucurbits, tomatoes, and damping-off, root rot, crown rot, and stem rot of sweet and hot peppers, and other vegetables. An unidentified *Fusarium* species was the main cause of the "pecky rice," a serious diseases of rice panicles and seeds in some fields in the Surkhandarya region of Uzbekistan. However, the number of surveys conducted in the country still was few, and the species identifications were based mainly on morphological characteristics of their colonies and conidia, and, in some cases, also on the pathogenicity to host plants. Undoubtedly, in further studies, the number of detected diseases on various crops caused by *Fusarium* species will grow significantly. This process will be particularly facilitated by the fact that in recent years, methods based on the DNA analysis started to be used to identify plant pathogenic fungi, especially *Fusarium* species, along with the classical methods.

Conflict of interest

The authors declare no conflict of interest. These authors contributed equally.

Abbreviations

BLAST	Basic Local Alignment Search Tool
CGB ASU	Centre for Genomics and Bioinformatics of the Academy of Sciences of the Republic of Uzbekistan
FCR/FRR/FHB	Fusarium crown rot/Fusarium root rot/Fusarium head blight
IGPEB ASU	Institute of Genetics and Plant Experimental Biology of the Academy of Sciences of the Republic of Uzbekistan
NCBI	National Center for Biotechnology Information, USA
PPS CRI	Plant Protection Station of the Cotton Research Institute, Uzbekistan
TSAU	Tashkent State Agrarian University, Uzbekistan

Author details

Batyr A. Khasanov[1*], Anvar G. Sherimbetov[2], Bakhtiyor S. Adilov[2]
and Albert A. Khakimov[1]

1 Tashkent State Agrarian University, Tashkent, Uzbekistan

2 Institute of Genetics and Plant Experimental Biology, Tashkent, Uzbekistan

*Address all correspondence to: khasanov.batyr@gmail.com

References

[1] Leslie JF, Summerell BA. The *Fusarium* Laboratory Manual. Ames, Iowa, USA: Blackwell Publishing; 2006. 388 pp. DOI: 10.1002/9780470278376

[2] Summerell BA, Salleh B, Leslie JF. A utilitarian approach to *Fusarium* identification. Plant Disease. 2003;**87**(2): 117-128

[3] Dean R, Van Kan JAL, Pretorius ZA, et al. The top 10 fungal pathogens in molecular plant pathology. Molecular Plant Pathology. 2012;**13**(4):414-430. DOI: 10.1111/J.1364-3703.2011.00783.X

[4] O'Donnell K, Ward TJ, Robert VARG, Crous PW, Geiser DM, Kang S. DNA sequence-based identification of *Fusarium*: Current status and future directions. Phytoparasitica. 2015;**43**(3):583-595. DOI: 10.1007/s12600-015-0484-z

[5] O'Donnell K, Sutton DA, Rinaldi MG, et al. Internet-accessible DNA sequence database for identifying *Fusaria* from human and animal infections. Journal of Clinical Microbiology. 2010;**48**: 3708-3718. DOI: 10.1128/JCM.00989-10

[6] O'Donnell K, Al-Hatmi AMS, Aoki T, et al. No to *Neocosmospora*: Phylogenomic and practical reasons for continued inclusion of the *Fusarium solani* species complex in the genus *Fusarium*. mSphere. 2020;**5**(5) 1-7: e00810-e00820. DOI: 10.1128/mSphere.00810-20

[7] Homa M, Galgóczy L, Manikandan P, et al. South Indian isolates of the *Fusarium solani* species complex from clinical and environmental samples: Identification, antifungal susceptibilities, and virulence. Frontiers in Microbiology. 2018;**9**:1-13. 1052. DOI: 10.3389/fmicb.2018.01052

[8] Summerell BA, Laurence MH, Liew ECY, Leslie JF. Biogeography and phylogeography of *Fusarium*: A review. Fungal Diversity. 2010;**44**:3-13. DOI: 10.1007/s1325-010-0060-20

[9] Taylor JW, Jacobson DJ, Kroken S, et al. Phylogenetic species recognition and species concepts in fungi. Fungal Genetics and Biology. 2000;**31**:21-32. DOI: 10.1006/fgbi.2000.1228

[10] O'Donnell K, Whitaker BK, Laraba I, et al. DNA sequence-based identification of *Fusarium*: A work in progress. Plant Disease. 2022;**106**(6): 1597-1609. DOI: 10.1094/PDIS-09-21-2035-SR

[11] Torres-Cruz TJ, Whitaker B, Proctor RH, et al. FUSARIUM-ID v. 3.0: An updated, downloadable resource for *Fusarium* species identification. Plant Disease. 2022;**106**(6):1610-1616. DOI: 10.1094/PDIS-09-21-2105-SR

[12] Schoch CL, Seifert KA, Huhndorf S, et al. Nuclear ribosomal internal transcribed spacer (ITS) region as a universal DNA barcode marker for fungi. Proceedings of the National Academy of Sciences USA (PNAS). 2012; **109**:6241-6246. DOI: 10.1073/pnas.1117018109

[13] Geiser DM, Jimenez-Gasco M, Kang S, et al. FUSARIUM-ID v. 1.0: A DNA sequence database for identifying *Fusarium*. European Journal of Plant Pathology. 2004;**110**:473-479. DOI: 10.1023/B:EJPP.0000032386.75915.a0

[14] Laraba I, McCormick SP, Vaughan MM, Geiser DM, O'Donnell K. Phylogenetic diversity, trichothecene potential, and pathogenicity within

Fusarium sambucinum species complex. PLoS One. 2021;**16**(1): e0245037. DOI: 10.1371/journal. pone.0245037

[15] Laraba I, Busman M, Geiser DM, O'Donnell K. Phylogenetic diversity and mycotoxin potential of emergent phytopathogens within the *Fusarium tricinctum* species complex. Phytopathology. 2022;**112**(6):1284-1298 Available from: https://apsjournals.apsnet.org/doi/10.1094/PHYTO-09-21-0394-R?ai=ru&ui=3x4f&af=T

[16] Leslie JF, Summerell BA. An overview of *Fusarium*. In: Brown DW, Proctor RH, editors. *Fusarium*. Genomics, Molecular and Cellular Biology. UK: Caister Academic Press; 2013. pp. 1-9

[17] FAOSTAT. 2023. Available from: https://www.fao.org/faostat/en/#data/QCL [Accessed: July 27, 2023]

[18] Khasanov BA. Imperfect Fungi as Main Causal Agents of Cereal Diseases in Middle Asia and Kazakhstan [Thesis]. Moscow: Moscow State University; 1992 (In Russian)

[19] Turakulov KS, Baboev SK, Gulmurodov RA. Rust Diseases of Wheat. Tashkent: Publishing House "Navroz"; 2015. p. 116

[20] Gulmurodov RA. Wheat Diseases in the Central and Southern Regions of Uzbekistan and their Control [Thesis]. Tashkent: Tashkent State Agrarian University; 2017 (in Uzbek)

[21] Khaytbaeva NS. Fusarium Diseases of Wheat in Saline Soils of the Republic of Karakalpakstan and Measures on their Control [Abstract of Thesis]. Tashkent: Tashkent State Agrarian University; 2017 (In Uzbek and Russian, with English summary)

[22] Khasanov BA, Turdieva DT, Safarov AA, Tukhtamyshev SS. Root Rot Diseases of Wheat. Tashkent: Publishing House of the Tashkent State Agrarian University; 2020. p. 144 (in Uzbek, with Russian and English abstracts)

[23] Khasanov BA, Safarov AA, Turdieva DT. Fusarium root and foot diseases of wheat in the world and Uzbekistan (review). Uzbek Biological Journal. 2020;**5**:21-32 (in Russian, with Uzbek and English abstracts)

[24] Khasanov BA, Gulmurodov RA, Sherimbetov AG. Fusarium head blight of wheat (review). Bulletin of the Uzbek National University. 2020;**3**(2):91-96 (in Russian, with Uzbek and English abstracts)

[25] Sheraliev A. Genus *Fusarium* Lk. Et Fr. *i*n Uzbekistan (Systematics, Distribution, and Biology) [Thesis]. Tashkent: Tashkent State Agrarian University; 2001 (in Uzbek)

[26] Baygulova GK, Goldschtein LE, Ellanskaya IA. Fusarium diseases of wheat on dryland ("boghara") areas of Uzbekistan. Uzbek Biological Journal. 1975;**2**:77-78 (in Russian)

[27] Goldschtein LE, Baygulova GK. Wheat root rot diseases in dryland ("boghara") areas of Uzbekistan. Mikologiya i Fitopatologiya. 1972;**6**(1): 524-528 (in Russian)

[28] Sherimbetov AG. Biomorphology and Pathogenic Properties of the Fungi from the Genus *Fusarium* [Thesis]. Tashkent: Institute of Microbiology, Uzbek Academy of Sci; 2019 (in Uzbek)

[29] Turdieva DT, Sherimbetov AG, Zokirov SS, Khasanov BA. Root rot diseases of winter wheat in Uzbekistan. Uzbek Biological Journal. 2019;**5**:18-25

(in Russian, with Uzbek and English summaries)

[30] Khasanov BA, Turdieva DT, Sherimbetov AG. Root and foot rot diseases of winter wheat in Uzbekistan. EPRA International Journal of Research and Development. 2019;**4**(9):21-26

[31] Bilai VI. The *Fusaria*. 2nd ed. Kiev, USSR: Naukova Dumka; 1977. p. 443 (in Russian)

[32] Sandoval-Denis M, Lombard L, Crous PW. Back to the roots: A reappraisal of *Neocosmospora*. Persoonia. 2019;**43**:90-185. DOI: 10.3767/persoonia.2019.43.04

[33] Seifert KA, Aoki T, Baayen RP, et al. The name *Fusarium moniliforme* should no longer be used. Mycological Research. 2003;**107**:634-644. DOI: 10.1017/S095375620323820X

[34] Gräfenhan T, Schroers H-J, Nirenberg HI, Seifert KA. An overview of the taxonomy, phylogeny, and typification of nectriaceous fungi in *Cosmospora, Acremonium, Fusarium, Stilbella*, and *Volutella*. Studies in Mycology. 2011;**68**:79-113. DOI: 10.3114/sim.2011.68.04

[35] Cook RJ. Fusarium root, crown, and root rots and associated seedling diseases. In: Bockus WW, Bowden RL, Hunger RM, Morrill WL, Murray TD, Smiley RW, editors. Compendium of Wheat Diseases and Pests. 3rd ed. MN, USA: APS Press; 2010. pp. 37-39

[36] Summerell BA, Leslie JF, Liew ECY, et al. *Fusarium* species associated with plants in Australia. Fungal Diversity. 2011;**46**(1):1-27. DOI: 10.1007/s13225-010-0075-8

[37] Obanor F, Chakraborty S. Aetiology and toxigenicity of *Fusarium graminearum* and *Fusarium pseudograminearum* causing crown rot and head blight in Australia under natural and artificial infections. Plant Pathology. 2014;**63**(6):1218-1229. Available from: https://bsppjournals.onlinelibrary.wiley.com/doi/epdf/10.1111/ppa.12200

[38] Xu F, Yang G, Wang J, Song Y, Liu L, Zhao K, et al. Spatial distribution of root and crown rot fungi associated with winter wheat in the North China plain and its relationship with climate change. Frontiers in Microbiology. 2018; **9**:1054-1064. 1064. DOI: 10.3389/fmicb.2018.01054

[39] Minati MH. First record of nine *Fusarium* species causing root rot on wheat (*Triticum aestivum* L.) in Iraq. In: AIP Conference Proceedings 2290, 020009. Karbala, Iraq: AIP (American Institute of Physics) Publishing; 2020. p. 18. DOI: 10.1063/5.0027398. Published Online: December 04, 2020

[40] Shrestha S, Poudel RS, Zhong S. Identification of fungal species associated with crown and root rots of wheat and evaluation of plant reactions to the pathogens in North Dakota. Plant Disease. 2021;**105**(11):3564-3572. DOI: 10.1094/PDIS-11-20-2412-RE?ai=rs&ui=3x4f&af=T

[41] Tunali B, Nicol JM, Hodson D, et al. Root and crown rot fungi associated with spring, facultative, and winter wheat in Turkey. Plant Disease. 2008;**92**(9): 1299-1306. DOI: 10.1094/PDIS-92-9-1299

[42] Bentley AR, Tunali B, Nicol JM, Burgess LW, Summerell BA. A survey of *Fusarium* species associated with wheat and grass stem bases in northern Turkey. Sydowia. 2006;**58**(2):163-177 Available

from: https://www.zobodat.at/publikation_articles.php?id=126956

[43] Chakraborty S, Duveiller E. Fusarium head blight and its relation to crown rot. In: Nicol JM, Bentley AR, Ferrar PJ, editors. Soilborne Pathogens of Wheat: Their Biology, Economic Importance and Integrated Control. 4th International Master Class in soil borne pathogens of wheat. Advanced theoretical training manual. Vol. 2010. Turkey: Anadolu Research Institute; 2010. pp. 51-58

[44] Dill-Macky R. Fusarium head blight (scab). In: Bockus WW, Bowden RL, Hunger RM, Morrill WL, Murray TD, Smiley RW, editors. Compendium of Wheat Diseases and Pests. 3rd ed. MN, USA: APS Press; 2010. pp. 34-36

[45] Balmas V, Scherm B, Marcello A, et al. *Fusarium* species and chemotypes associated with Fusarium head blight and Fusarium root rot on wheat in Sardinia. Plant Pathology. 2015;**64**: 972-979. DOI: 10.1111/ppa.12337. Published online 8.01.2015

[46] Lee T, Zhang H, Diepeningen A, Waalwijk C. Biogeography of *Fusarium graminearum* species complex and chemotypes: A review. Food Additives & Contaminants, Part A. 2015;**32**(4): 453-460. DOI: 10.1080/19440049.2014.984244n

[47] Knight NL, Sutherland MW. Assessment of *Fusarium pseudograminearum* and *F. Culmorum* biomass in seedlings of potential host cereal species. Plant Disease. 2017; **101**(12):2116-2122. DOI: 10.1094/PDIS-12-16-1739-RE

[48] Bissonnette K, Wharton P, Chen J, Marshall JM. Survey of *Fusarium* species associated with Fusarium head blight of spring wheat (*Triticum aestivum*) in southern Idaho. Plant Health Progress. 2018;**19**(2):125-127. DOI: 10.1094/PHP-10-17-0066-S

[49] Haile JK, N'Diaye A, Walkowiak S, et al. Fusarium head blight in durum wheat: Recent status, breeding directions, and future research prospects. Phytopathology. 2019;**10**(10): 1664-1675. DOI: 10.1094/PHYTO-03-19-0095-RVW

[50] Valverde-Bogantes E, Bianchini A, Herr JR, et al. Recent population changes of Fusarium head blight pathogens: Drivers and implications. Canadian Journal of Plant Pathology. 2019, 2019; **42**:315-329. DOI: 10.1080/07060661.2019.1680442

[51] Chełkowski J, editor. Fusarium Mycotoxins, Taxonomy and Pathogenicity. Amsterdam – Oxford – New York – Tokyo: Elsevier; 1989. p. 492. DOI: 10.3390/microorganisms8091404

[52] Egamberdiev SS. Molecular Identification and Characterization of Pathogens from the Genus *Fusarium* Affecting Cotton in Uzbekistan [Thesis]. Tashkent: Institute of Gene Pool of Plants and Animals, National University of Uzbekistan, and Institute of Genetics and Plant Experimental Biology; 2016 (in Russian)

[53] Moparthi S, Burrows M, Mgbechi-Ezeri J, Agindotan B. *Fusarium* spp. associated with root rot of pulse crops and their cross-pathogenicity to cereal crops in Montana. Plant Disease. 2021; **105**(3):548-557 Available from: https://apsjournals.apsnet.org/doi/10.1094/PDIS-04-20-0800-RE?ai=rs&ui=3x4f&af=T

[54] Gagkaeva T, Gavrilova O, Levitin M, Novojilov K. Fusarium diseases of cereals. Addendum to the Journal "Zaschita i karantin rasteniy" (Moscow). 2011;**5**:75-120 (in Russian)

[55] Wagacha JM, Steiner U, Dehne H-W, et al. Diversity in mycotoxin and fungal species infecting wheat in Nakuri district. Kenya Journal of Phytopathology. 2010;**158**(7–8):527-535. DOI: 10.1111/j.1439-0434.2009.01653.x

[56] Eslahi MR. Fungi associated with root and crown rot of wheat in Khuzestan. Iranian Journal of Crop Protection. 2012;**1**(2):107-113 Available from: https://jcp.modares.ac.ir/files/jcp/user_files_749497/archive_global-A-3-1000-8804-81ec25a.pdf

[57] Kebede M, Adugna G, Hundie B. Identification of *Fusarium* species responsible to cause wheat head blight in southwestern Ethiopia. Research Journal of Plant Pathology. 2020;**3**(1:04):1-8. DOI: 10.36648/plantpathology.3.1.04

[58] Yang M, Zhang H, Kong X, et al. Host and cropping system shape the *Fusarium* population: 3ADON-producers are ubiquitous in wheat whereas NIV-producers are more prevalent in rice. Toxins. 2018;**10**(115):1-12. DOI: 10.3390/toxins10030115

[59] Gagkaeva TY, Gavrilova OP, Orina AS. First record of *Fusarium globosum* in the mycobiota of cereals on the territory of Ural and Siberia. Vestnik Zacshiti Rasteniy. 2009;**1**(99):10-18 (in Russian. DOI: 10.31993/2308-6459-2019-1(99)-10-18

[60] Wallwork H. Cereal Root and Crown Diseases. Canberra, Australia: Grains Research Development Corporation Publications; 2000. p. 62

[61] Burgess LW, Bentley AR, Wallwork H. Crown rot of wheat. In: Nicol JM, Bentley AR, Ferrar PJ, editors. Soilborne Pathogens of Wheat: Their Biology, Economic Importance and Integrated Control. 4th Int. Master Class in soilborne pathogens of wheat. Advanced theoretical training manual. Turkey: Anadolu Res. Inst; 2010. pp. 45-50

[62] Castañares E, Stenglein SA, Dinolfo MI, Moreno MV. *Fusarium tricinctum* associated with head blight on wheat in Argentina. Plant Disease. 2011; **95**(4):496. DOI: 10.1094/PDIS-07-10-0485

[63] Martínez M, Dinolfo MI, Nogueira S, Stenglein SA. *Fusarium tricinctum* associated with head blight on barley in Argentina: Pathogenicity and potential degradation of the different hordein fractions. Journal of Plant Diseases and Protection. 2021;**128**:1377-1381. DOI: 10.1007/s41348-021-00459-6

[64] Wang Y, Wang R, Sha Y. Distribution, pathogenicity and disease control of *Fusarium tricinctum*. Frontiers in Microbiology. 2022;**13**(939927):1-11. DOI: 10.3389/fmicb.2022.939927

[65] Lombard L, Merwe NA, Groenewald JZ, Crous PW. Generic concepts in Nectriaceae. Studies in Mycology. 2015;**80**:189-245. DOI: 10.1016/j.simyco.2014.12.002

[66] Abdallah-Nekache N, Laraba I, Ducos C, et al. Occurrence of Fusarium head blight and Fusarium crown rot in Algerian wheat: Identification of associated species and assessment of aggressiveness. European Journal of Plant Pathology. 2019;**154**(3):499-512. DOI: 10.1007/s10658-019-01673-7

[67] Solov'yova AI, Poyarkova LV. Cotton Wilt. Tashkent: Publishing House "Sel'khozgiz of Uzbek SSR"; 1940. p. 64. (in Russian)

[68] Solovyova AI, Poyarkova LV. Fusarium Wilt of the Egyptian Cotton. Tashkent: Publishing House "Gosizdat of Uzbek SSR"; 1945. 88 p. (in Russian)

[69] Solovyova AI. Cotton Wilt (Abstract of Thesis). Tashkent: Sredazgidroenergostroy; 1954. p. 34. (in Russian)

[70] Matveev GG. Fusarium wilt of upland cotton. Mikologiya I Fitopatologiya. 1984;**18**(5):412-415 (in Russian)

[71] Edel-Hermann V, Lecomte C. Current status of *Fusarium oxysporum* formae speciales and races. Phytopathology. 2019;**109**(4):512-532. DOI: 10.1094/PHYTO-08-0320-RVW

[72] Jonavičienė A, Supronienė S, Semaškienė R. *Microdochium nivale* and *M. majus* as causative agents of seedling blight in spring cereals. Zemdirbyste-Agriculture. 2016;**103**(4):363-368. DOI: 10.13080/z-a.2016.103.046

[73] O'Donnell K, McCormick SP, Busman M, et al. Marasas et al. 1984 "toxigenic *Fusarium* species: Identity and mycotoxicology" revisited. Mycologia. 2018;**110**(6):1-23. DOI: 10.1080/00275514.2018.1519773

[74] Marupov A, Ishankulova M, Rahmatov A. New causal agent of cotton Fusarium wilt. Journal "Sel'skoe khozyaystvo Uzbekistana". 2008;**5**:30 (in Russian)

[75] Khasanov BA. Fusarium wilt of cotton: Can *Fusarium moniliforme* Sheldon be a cause? Turkish Soil-Water Journal. 2013;**2**(2/2):2299-2306

[76] Khasanov BA. Fusarium Wilt of Cotton and Current Methods of Identification of *Fusarium* Species. Tashkent: Publishing House of the Tashkent State Agrarian University; 2017. p. 136. (in Russian)

[77] Marupov A, Stipanovic RD, Turamuratova GH, Mambetnazarov AB, Marupova MA. *Fusarium verticillioides*: A new cotton wilt pathogen in Uzbekistan. International Journal of Plant Disease and Pathology. 2013;**1**(1):1-5 Available from: http://acascipub.com/Journals.php

[78] Egamberdiev SS, Ulloa M, Saha S, et al. Molecular characterization of Uzbekistan isolates of *Fusarium oxysporum* f. sp. vasinfectum. Journal of Plant Science and Molecular Breeding. 2013;**2**(1):3. DOI: 10.7243/2050-2389-2-3

[79] Radjapov F, Egamberiev S, Salakhutdinov I, et al. The use of single-copy genes for the identification of phytopathogens of the genus *Fusarium* in Uzbekistan. In: Materials of the Republican Scientific-Practical Conference "Important Directions of the Organization of Scientific Research on Breeding and Seed Production". Tashkent: Tashkent State Agrarian University; 2013. pp. 293-295 (in Russian)

[80] Booth C. The genus *Fusarium*. Kew, Surrey, England: The Commonwealth Mycological Institute; 1971. p. 858

[81] Sheraliev A. Fusarium Diseases of Mulberry Trees. Tashkent: Publishing house "Fan"; 1992. p. 108 (in Uzbek)

[82] Mirpulatova NS. Fusarium wilt of melons in Uzbekistan. In: Pests and Diseases of Cotton and Other Crops, and their Control. Collection of Scientific Papers. Tashkent: All-Union Scientific Research Institute of Cotton Growing; 1951. pp. 204-210 (in Russian)

[83] Khasanov BA, Ochilov RO, Gulmurodov RA. Diseases of Vegetables, Potato and Cucurbit Crops, and their Control. Tashkent: Publishing house "Voris-Nashriyot"; 2009. p. 245. (in Uzbek, with Russian and English summaries)

[84] Pestsova ST, Borisov NS. Infection of tomato varieties with Fusarium diseases. In: "Control of Pests, Diseases, and Weeds of Plants". Collection of Scientific Papers. Tashkent: Publishing house "TIPO after Ibn Sino"; 1995. pp. 51-53 (in Uzbek)

[85] Pestsov VI. The content of *Fusarium* species in vegetable and melon crops. In: Materials of the Jubilee Republican Conference on Microbiology, Algology and Mycology, Dedicated to the 50th Anniversary of the Uzbek SSR and the Communist Party of Uzbekistan. Tashkent: Fan; 1974. pp. 178-179 (in Russian)

[86] Khakimov AA, Khasanov BA, Khamiraev UQ, et al. Fusarium diseases of capsicums. Journal "Agrokimyohimoya and Plant Quarantine" (Uzbekistan). 2021;**4**:72-77 (in Uzbek, with Russian and English abstracts)

[87] Khasanov BA, Khakimov AA, Khamiraev UQ, Utaganov SB, Aznabakieva DT. Genus *capsicum* and diseases of sweet and hot peppers (review). Bulletin of Science and Practice. 2021;**7**(10):98-114. DOI: 10.33619/2414-2948/71/12 (In Russian, with English abstract)

[88] Khasanov BA, Khakimov AA, Aznabakieva DT, Khamiraev UQ, Utaganov SB. Fusarium diseases of sweet and hot peppers (an overview). Uzbek Biological Journal. 2022;**1**:33-43 (in Russian, with Uzbek and English abstracts)

[89] Khasanov BA. An overview of infectious diseases of rice registered in Uzbekistan. Uzbek Biological Journal. 2010;**6**:20-24 (in Russian, with Uzbek and English abstracts)

[90] Akbarov MM, Khakimova NT. The significance of green and globe (summer) radishes and their fungal diseases. Journal "Agrokimyohimoya and Plant Quarantine" (Uzbekistan). 2023;**2**(Special issue):216-220 (in Uzbek)

[91] Groth D, Hollier C, Rush C. 2009. Rice disease identification power point. Photo Link. LSU AgCenter, Department of Plant Pathology and Crop Physiology. Baton Rouge LA. Available from: www.agri971.yolasite.com. https://www.lsuagcenter.com/topics/crops/rice/diseases/rice-disease-identification-powerpoint

[92] Groth DE, Rush MC, Hollier CA. Rice Diseases and Disorders in Louisiana. LSU Agricultural Experiment Station Reports. Vol. 668. Bulletin No. 828. Louisiana State University Agricultural Center; 1991. Available from: http://digitalcommons.lsu.edu/agexp/668

[93] Nikiema FW, Zida EP, Keita I, et al. Molecular identification of *Fusarium* species associated with rice seeds produced in Burkina Faso. Journal of Agriculture and Environmental Sciences. 2023;**12**(1):6-17. DOI: 10.15640/jaes.v12n12a2

[94] Lomas-Cano T, Boix-Ruiz A, García-Rodríguez C, et al. Etiological and epidemiological concerns about pepper root and lower stem rot caused by *Fusarium oxysporum* f. sp. *radicis-capsici* f. sp. nova. Phytoparasitica. 2016;**44**(3): 283-293. DOI: 10.1007/s12600-016-0522-5

[95] Lomas-Cano T, Palmero-Llamas D, Cara M, et al. First report of *Fusarium oxysporum* on sweet pepper seedlings in Almería, Spain. Plant Disease. 2014; **98**(10):1435. DOI: 10.1094/PDIS-04-14-0365-PDN

[96] Cerkauskas R. Fusarium Stem and Fruit Rot of Greenhouse Pepper. Factsheet 294/638. Ontario, Canada:

OMAFRA (Ontario Ministry of Agriculture, Food and Rural Affairs); 2001. Available from: http://www.omafra.gov.on.ca/english/crops/facts/01-083.htm [Accessed: August, 2009]

[97] Bouanaka H, Bellil I, Khelifi D. *Fusarium brachygibbosum*, identification and virulence as causal agents of crown rot on wheat. Revue Agrobiologia. 2023; **13**(1):3361-3372. Available from: https://www.asjp.cerist.dz/en/article/228544

[98] Lin F, Chhapekar SS, Vieira CC, et al. Breeding for disease resistance in soybean: A global perspective. Theoretical and Applied Genetics. 2022; **135**:3773-3872. DOI: 10.1007/s00122-022-04101-3

[99] Xia JW, Sandoval-Denis M, Crous PW, Zhang XG, Lombard L. Numbers to names – Restyling the *Fusarium incarnatum-equiseti* species complex. Persoonia. 2019;**43**:186-221. DOI: 10.3767/persoonia.2019.43.05

Chapter 3

Early Selection for Resistance to *Fusarium* Wilt in Banana

Barbarita Companioni González, Rómulo García Velasco, José Carlos Lorenzo Feijoo and Ramón Santos Bermudez

Abstract

Early selection for plant resistance to different special forms of *Fusarium oxysporum* has been a key goal in conventional and biotechnological breeding. We previously developed a straightforward procedure to differentiate resistant and susceptible field-grown banana cultivars for *Fusarium* wilt at the leaf level. This chapter presents compiled results from several years of work by our group, developing a method to differentiate resistance and susceptibility to *Fusarium* wilt vegetative compatibility group (VCG) [01210] race 1 at the leaf level in banana cultivars. This is achieved using the leaf pit bioassay and application of fungal culture filtrates.

Keywords: bioassay, biotechnology, disease, fungus filtrates, *Musa* spp.

1. Introduction

Bananas and plantains, belonging to the *Musa* spp., stand as pivotal crops in tropical and subtropical regions. They serve as a crucial food source for a substantial portion of the global population [1]. However, the production of these crops faces a significant threat from diseases like *Fusarium* wilt, primarily caused by *Fusarium oxysporum* f. sp. *cubense* (FOC) [2]. The introduction of FOC tropical race 4 (RT4) into Cavendish cultivar plantations has marked it as one of the top ten diseases in agricultural history [3, 4]. Controlling this disease with agrochemicals comes at a high cost and inflicts severe environmental damage [1]. As a result, there has been a growing consensus in recent years that the most effective and environmentally friendly approach to combat this disease is genetic enhancement for resistance [5].

In this context, the initial and pivotal step in assessing a new banana clone within a breeding program for this crop is selecting for resistance to different races of FOC wilt [6]. Simultaneously, early selection for plant resistance against various specialized forms (f. sp.) of *F. oxysporum* has been a primary focus in conventional and biotechnological breeding. Advancements in our understanding of plant processes, pathogen biology, and plant-pathogen interactions have bolstered this effort. Nevertheless, existing methods for screening resistance to this disease under field conditions are labor-intensive, destructive, and time-consuming [7, 8]. Consequently, developing efficient techniques for selecting susceptible and resistant cultivars to combat this pathogen in crop breeding has become a top priority.

Hence, the utilization of fungal culture filtrates (FCH), toxins, and even dual cultures holds promise in this endeavor. This chapter presents the results of several years of research conducted by our working group in developing a method for distinguishing resistance and susceptibility to *Fusarium* wilt race 1 (VCG) [01210] in bananas at the leaf level, employing FCH and the puncture bioassay in an efficient and non-destructive manner.

2. Early resistance selection

For many years, assessing plant disease resistance has primarily relied on *ex vitro* and *in vivo* selection. These approaches involve exposing the plants to natural infection by fungus spores and continue to be widely employed even today. The results of evaluating resistance to FOC's RT4 strain in banana cultivars, plantains, and Cavendish somaclonal variants using these selection techniques, as provided by the Taiwan Banana Research Institute, are presented in **Table 1** [9].

Li et al. [14] assessed the resistance of eight wild banana genotypes (*Musa acuminata* subsp. *burmannica*, *M. balbisiana*, *M. basjoo*, *M. itinerans*, *M. nagensium*, *M. ruiliensis*, *M. velutina*, and *M. yunnanensis*) to FOC's RT4 strain under both greenhouse and field conditions. In the greenhouse experiments, banana seedlings from *in vitro* cultures were infected by puncturing holes at the base of the pseudo stem with a fungus spore suspension of $5{\times}10^6$ spores mL^{-1}. They then monitored resistance every 2 days for 65 days by assessing the seedlings' internal and external disease symptoms. External symptoms were evaluated using a four-class rating scale: 0 = no symptoms; 1 = initial yellowing, mainly on lower leaves; 2 = yellowing of all lower leaves with some discoloration of younger leaves; 3 = all leaves with intense yellowing or dead plant. Internal symptoms were assessed based on rhizome discoloration:

Cultivars	Place	Response to FOC RT4	Reference
'FHIA 01'	FAO, Malasia	Resistant	[9]
'FHIA 02'	International *Musa* Testing Program, phase III, China (IMTP III, China)	Resistant	[9]
'FHIA 03'	Journal of Tropical Crops China	Resistant	[10]
'FHIA 17'	Papua New Guinea	Resistant	[11]
'FHIA 18'	Journal of Tropical Crops China	Highly resistant	[10]
'FHIA 21'	International *Musa* Testing Program, phase III, China (IMTP III, China)	Resistant	[11]
'FHIA 23'	FHIA: Bananas Database	Susceptible	[12]
'FHIA 25'	Papua New Guinea	Resistant	[11]
'FORMOSANA'	Taiwan	Resistant	[13]

Table 1.
Results in the evaluation for resistance to RT4 of FOC in banana, plantains, and somaclonal variants of Cavendish cultivars.

0 = no symptoms, 1 = 1–20%, 2 = 21–40%, and 3 = > 40% discolored rhizome. They conducted experiments in a plot naturally infested with FOC RT4 for the field screening test. They evaluated external disease symptom expression every 2 weeks for 12 months after planting, and assessed internal symptoms at the end of the 12-month experiment using the same scale. The results demonstrated the presence of various sources of resistance to FOC's RT4, representing valuable genetic resources for banana breeding programs aiming to develop wilt-resistant cultivars against this strain. However, research on resistance to RT4 of FOC continues to be an ongoing endeavor.

In a similar approach [5] screened banana mutants from the Rasthali cultivar (Silk, AAB) were resistant to GCV race 1 [0124/5] of FOC through *in vitro* selection. They employed FCH (fungal culture filtrates) and the toxin fusaric acid as selection agents. They initiated the process by transferring individual banana mutant shoots, obtained after the third or fourth subculture, to a multiplication culture medium. This medium was enriched with varying concentrations of the toxin (commercial fusaric acid from Sigma Aldrich, USA) ranging from 0.0125 to 0.0625 mM, along with FCH concentrations spanning from 3–8%. Additionally, a growth regulator was introduced into the medium. After 3 weeks, they observed that 50% of the explants survived at the 0.050 mM toxin concentration and 7% with FCH. Higher concentrations resulted in a noticeable decline in the shoot growth of the banana mutants. The selected mutants were transferred to pot tests under controlled conditions 3 months later. The substrate was inoculated with a spore suspension containing 12 × 109 conidia mL^{-1} of the fungus to initiate these tests. After 6 months of evaluating resistance, they successfully identified three potential mutants that exhibited resistance to race 1 of FOC. These promising mutants were subsequently propagated extensively *in vitro* to enable further studies on their interaction with the fungus.

It's important to note that the selection agent chosen plays a crucial role in establishing resistance selection systems. To achieve this, determining the correct inoculum concentration of the FCH toxin is essential to highlight the differential phytotoxic effects among different varieties. This step significantly enhances the likelihood of obtaining stable plant lines with disease resistance [15]. Moreover, even when *in vitro* selections are conducted, subjecting the supposedly resistant plants to field studies is imperative. It's worth mentioning that the disease caused by FOC exhibits relatively long incubation periods under these conditions. In simpler terms, it may take several months from root infection to the appearance of external symptoms like yellowing and leaf blade collapse. Consequently, monitoring a large number of plants over an extended period becomes necessary.

These approaches underscore the urgency of developing alternative, swift, and non-destructive methods for screening banana resistance to FOC in resistance screening tests conducted under field and *ex vitro* conditions.

2.1 Use of fungal culture filtrates for leaf-level differentiation of resistance to *Fusarium* wilt race 1 (VCG) [01210]

Considering the earlier approaches, [16] devised a method to distinguish between resistance and susceptibility of banana cultivars to FOC race 1, GCV [01210]. They achieved this using a leaf spotting bioassay combined with applying FCH. The assessment involved gathering middle-aged leaves from two banana cultivars: Gros Michel (group AAA, susceptible) and FHIA-01 (group AAAB, resistant). These leaves were then subjected to FCH application at three distinct positions on the upper side of the leaf blade: distal, middle, and proximal. After 48 hours, they evaluated the phytotoxic

A= ¶/4 ab A – ellipse area (mm^2) a – semi-major axis (mm) b – semi-minor axis (mm) ¶ - 3.14	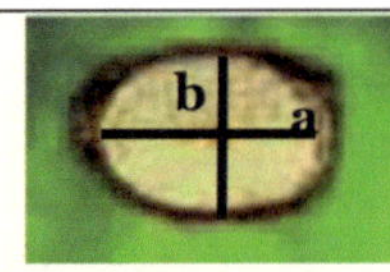

Table 2.
Symptom expression of necrosis formed around the point of application of the fungal culture filtrate, expressed as the area of the elliptical lesion (mm^2) [17].

effects of the FCH by examining the symptomatological manifestation of necrosis around the area where FCH was applied, quantified as the size of the elliptical lesion (mm^2) (**Table 2**) [17]. The most significant differences between cultivars were observed 48 hours after applying FCH to banana leaves.

In previous studies [18, 19] additional evaluations, including the analysis of biochemical components were conducted. The steps involved in this analysis are presented straightforwardly. Notably, a novel aspect of this research is the analysis to differentiate between resistance and susceptibility of banana cultivars within the genetic improvement program. This estimation was derived from a data matrix encompassing the impact of FCH, including factors such as lesion area, levels of free and cell wall-bound phenols, aldehydes (excluding malondialdehyde), and proteins in leaves from eighteen cultivars. Following the discriminant analysis, two functions were derived—one for resistant cultivars and another for susceptible cultivars. Each plant in the matrix underwent assessment based on these discriminant functions, resulting in a classification as resistant or susceptible in 94.4% of the cases (68 out of 72 plants). This described method proves to be a valuable tool for enhancing selection efficiency under both *ex vitro* and *in vivo* conditions, irrespective of environmental factors and the time of year conducive to disease development. Moreover, it facilitates evaluating a substantial number of samples in a laboratory setting, thus expediting genetic improvement programs for combatting this disease.

The described method introduces the possibility of screening for resistance to FOC in leaf fragments of plants aged between 8 to 9 months. In this approach, FCH is applied to the leaves of individuals under study for resistance or susceptibility, while a control group receives a non-inoculated culture medium. Comparing this selection method with existing techniques for evaluating banana resistance to *Fusarium oxysporum* f. sp. *cubense* (as previously developed by [20, 21], it becomes evident that working with FCH is notably more straightforward. Furthermore, this method offers advantages such as reducing the space requirements for the selection process and shortening the time needed to determine whether a plant is resistant or susceptible. Additionally, it addresses a drawback of resistance evaluations conducted under *ex vitro* conditions, which often prove ineffective due to the brief infection period and challenges associated with controlling inoculum concentration distribution. It also mitigates the issues with resistance selection under natural field conditions, which tend to be time-consuming, costly, and influenced by fluctuations in inoculum abundance and climatic factors impacting disease dissemination, infection, development, and symptom expression [22].

As previously mentioned, [16, 19] developed a method for distinguishing the resistance of banana cultivars to FOC race 1 using FCH. However, the broad application of these results in genetic improvement programs for this crop necessitates validation against different races of the fungus. Specifically, validation was required for two isolates belonging to different vegetative compatibility groups (GCV): GCV [01210] race 1 and GCV [0124/125] race 2. To address this, [23] undertook the validation of this method for the rapid differentiation of Panama disease using FCH from both FOC race 1 and race 2. For this validation, they utilized foliar samples from banana plants aged 8 to 9 months, cultivated at the Banana Germplasm Bank of the National Institute for Tropical Viandas Research in Santo Domingo, Villa Clara, Cuba. These banana cultivars had previously been classified as either resistant or susceptible to FOC races 1 and 2 by [24]. The fungal strains employed were FOC race 1 GCV [01210] and race 2 GCV [0124/125], sourced from the strain collection of the Instituto de Investigaciones de Sanidad Vegetal (INISAV), Havana, Cuba. In the initial part of the experiment, [23] focused on obtaining FCH during the *in vitro* growth of both FOC race 1 GCV [01210] and race 2 GCV [0124/125]. To achieve this, they determined the optimal harvesting time for FCH for both races on a daily basis for up to 30 days. The phytotoxic activity of the FCH was assessed using the puncture bioassay method outlined by [16] at each evaluated time point (ranging from 1 to 30 days). During this research stage, they compared the responses of 15 banana cultivars with varying resistance levels to the pathogen *in vivo*, as previously categorized by [24]. FCH for FOC race 1 was harvested at 15 days, while that for race 2 was collected at 29 days after inoculation in Caldo Czapek culture medium. Subsequently, they evaluated how cultivars of *Musa* spp., responded to the application of fungal filtrates from both races.

The FCH obtained at 15 and 29 days from strains of races 1 and 2 successfully differentiated over 93% of individuals with known responses to the pathogen *in vivo* (**Table 3**). Among the cultivars most susceptible to FCH from race 1 were Manzano Criollo, Gros Michel, Pisang Lilin, and Yangambi km 5, all showing no statistical differences in their responses. Paka followed as the next most susceptible cultivar. Conversely, all FHIA cultivars assessed (01, 02, 03, 04, 18, and 21) displayed resistance to the phytotoxic effects of FCH from race 1 GCV [01210], with no statistical variations in their respective responses. Furthermore, the cultivars 'Burro Criollo,' 'Pelipita,' 'Pisang jari guaya,' and 'Bluggoe' also exhibited resistance, with responses similar to FHIA. Researchers in various plant-pathogen interactions have successfully obtained culture filtrates with phytotoxic properties for early resistance screening. Moreover, methods have been developed for the isolation and purification of metabolites that play a role in the selective response of plants [25].

The results indicate that banana plants' resistance or susceptibility to *Fusarium oxysporum* f. sp. *cubense* can be distinguished at the leaf level using fungal culture filtrates, applicable in conventional and biotechnological *Musaceae* genetic improvement programs. In this context, one challenge in testing resistance under *in vivo* conditions is ensuring a uniform distribution of the pathogen inoculum among all plants to be evaluated. Some plants may receive excessive inoculum, while others might not get enough to develop the disease [26]. Additionally, environmental effects on inoculated plants studied in the field raise concerns [22]. Furthermore, it's not always feasible to characterize FOC populations through pathogenicity testing due to the influence of plant-environment interactions. There's clear evidence of differential plant responses under varying environmental conditions, even within established differential groups [27]. It's worth noting that to establish a disease resistance evaluation method, it must outperform the traditional approach in terms of labor, space, and time requirements. These considerations have been taken into account in the method tested for both pathogen races.

Cultivar	Elliptical lesion area (mm²)	
	FOC race 1 GCV [01210] (Day 15)	FOC race 2 GCV [0124/125] (Day 29)
Manzano Criollo (AAB)	***53.6a	*2.3b
Gros Michel (AAA)	***55.4a	*2.8b
Yangambi Km 5 (AAA)	***55.3a	*4.1b
Paka (AA)	***52.4a	*3.5b
Pisang Lilin (AA)	**54.4a	*3.4b
Pisang jari guaya (AA)	**4.1b	**52.8a
Burro Criollo (ABB)	*3.2b	***54.0a
Bluggoe (ABB)	*2.1b	***53.7a
Pelipita (ABB)	*3.4a	*2.6b
FHIA-01 (AAAB)	*3.6a	*2.3b
FHIA-02 (AAAB)	*2.5a	*2.6a
FHIA-03 (AABB)	**2.8b	***53.5a
FHIA-04 (AAAB)	*3.4a	*3.2a
FHIA-18 (AAAB)	**3.2b	*4.1a
FHIA-21 (AAAB)	*4.0b	**4.8a

*Means with equal letters indicate that there are no statistically significant differences (univariate analysis of variance, one-factor ANOVA, ANOVA, and Tukey's HSD, $p \leq 0.05$). Response of cultivars in field evaluations according to [24]: *Resistant, **Tolerant, ***Susceptible/Averages with same letters indicate no statistically significant differences (One-factor univariate analysis of variance, ANOVA, Tukey ANOVA, and HSD, $p \leq 0.05$). Responses of the cultivars in field evaluations according to [24]: *Resistant, **Tolerant, ***Susceptible.*

Table 3.
Responses of the Musa spp., cultivars to the application of filtrates of the fungus culture of races 1 and 2, evaluated with the bioassay on banana detached leaves and in earlier field evaluations faced with the microorganism in vivo [9].

The results from [23] highlight the feasibility of a rapid and non-destructive method for distinguishing susceptibility from resistance to *Fusarium* blight. This method utilizes FCH applied to leaves and can be employed for different pathogen populations, offering potential benefits for conventional and biotechnological *Musaceae* breeding programs. Moreover, these findings represent a pioneering step in validating a method that disrupts the natural biological cycle of the disease.

They also lay the groundwork for future research to identify phytotoxic metabolites and examine the specificity of putative avirulence genes in the pathogen. This, in turn, will facilitate an understanding of the role of these metabolites in the response of *Musa spp.*, to *Fusarium* wilt. In this context, [28] conducted a study using the bioassay described by Companioni *et al.* [16] to investigate the toxic components of FCHs from FOC GCV race 1 [01210]. They successfully isolated extracellular microbial metabolites involved in the plant's response, generating a host-specific culture filtrate. This research led to the characterization of phytotoxic compounds such as fusaric acid, bovericin, and fumonisin B1. These results lay the foundation for the direct isolation of avirulence genes in the pathogen and resistance genes in bananas, employing advanced genetics to enhance genetic improvement programs for the crop.

The results presented here underscore the viability of biotechnological techniques as valuable tools in the quest for FOC resistance. Nonetheless, many field studies lack the long-term data necessary to assess genes' effectiveness in real-world conditions. To address this, comprehending the host-pathogen interaction, particularly in terms of defense mechanisms and virulence pathways, becomes pivotal. Such understanding helps to pinpoint critical stages for developing resistant cultivars through genetic approaches. Nevertheless, selecting for resistance to this disease in field conditions remains a labor-intensive, destructive, and time-consuming process. This highlights the significance of developing efficient methods for early resistance selection. Such methods enable not only precise investigations into plant-pathogen interactions but also enhance the overall efficiency of breeding programs for this crop.

2.2 Evaluation of resistance to *Fusarium* wilt race 1 in banana mutants

The application of biotechnological techniques to bolster resistance breeding for *Fusarium* wilt in bananas has accelerated the introduction of breeding outcomes when contrasted with conventional methods [29]. Additionally, these techniques have enhanced the overall efficiency of breeding programs for this crop [5].

Ventura et al. [30] utilized a combination of tissue culture and *in vitro* mutagenesis in meristematic apices of the 'Zanzibar' cultivar (group AAB). This approach yielded a wide range of genetic variability in the 'Zanzibar' clone, leading to the selection of 15 mutants. Among these mutants, 'Z 13,' 'Z 30,' and 'Z 30 A' emerged as the most promising candidates for crop genetic improvement due to their impressive traits, including high yield, elimination of superficial corms, reduced height, and organized tillering. Given these promising findings, [31] assessed resistance to *Fusarium* wilt race 1 GCV [01210] in these selected banana clones. These clones were derived from the genetic improvement program for *Musa* spp., conducted at the Instituto Nacional de Investigaciones en Viandas Tropicales de Santo Domingo, Villa Clara, Cuba (INIVIT). These 15 mutants were initially obtained from the 'Zanzibar' clone through ionizing radiation (gamma rays) applied to meristematic apices in an *in vitro* culture. Following this, the most advanced materials for crop genetic improvement, specifically 'Z 13,' 'Z 30,' and 'Z 30 A' were chosen after three vegetative cycles under field conditions. These lines are being considered for registration as new genotypes in *Musa* spp., by the International Atomic Energy Agency (IAEA) [30]. To evaluate resistance to FOC race 1 GCV [01210], the methodology described by [18, 19] was employed. Foliar samples were collected from the 'Z 13,' 'Z 30,' and 'Z 30 A' lines 8 to 9 months after planting in the Banana Germplasm Bank of INIVIT. The evaluation of resistance to *Fusarium* wilt race 1 GCV [01210] in each line and at each assessment time, based on discriminant functions, is detailed in **Table 4**. After 48 hours following the application of treatments and subsequent biochemical analyses on the leaves of the 'Zanzibar' clone lines, it was observed that the discriminant functions classified 93.3% of 'Z 30' plants as resistant (56 out of 60 plants). Similarly, 91.6% of 'Z 13' plants were classified as resistant (55 out of 60 plants), and 'Z 30 A' exhibited a classification of 96.6% of plants as resistant (58 out of 60 plants). These results demonstrate that the selected lines derived from the irradiated 'Zanzibar' clone, namely 'Z 13,' 'Z 30,' and 'Z 30 A' hold substantial promise for genetic improvement in the crop, particularly for resistance to FOC race 1 GCV [01210], with resistance levels ranging from 91.6 to 96.6%. However, the necessity of continuing additional field studies to confirm resistance in the evaluated lines cannot be discounted. Even when *in vitro* resistance evaluations are conducted, it remains essential to subject putatively resistant plants to field studies [32].

Evaluation of moments	Evaluated lines	Total number of plants evaluated	Total number of potentially resistant plants according to discriminant functions
1	'Z-30'	10	10 (100%)
2	'Z-30'	10	8 (80%)
3	'Z-30'	10	9 (90%)
4	'Z-30'	10	9 (90%)
5	'Z-30'	10	10 (100%)
6	'Z-30'	10	10 (100%)
1	'Z-13'	10	8 (80%)
2	'Z-13'	10	9 (90%)
3	'Z-13'	10	9 (90%)
4	'Z-13'	10	10 (100%)
5	'Z-13'	10	10 (100%)
6	'Z-13'	10	9 (90%)
1	'Z-30A'	10	10 (100%)
2	'Z-30A'	10	9 (90%)
3	'Z-30A'	10	9 (90%)
4	'Z-30A'	10	10 (100%)
5	'Z-30A'	10	10 (100%)
6	'Z-30A'	10	10 (100%)

Table 4.
Classification into susceptible or resistant lines made by discriminant functions on selected lines of the irradiated 'Zanzíbar' clone [17].

The results obtained in this work showed that the selected lines of the irradiated 'Zanzibar' clone ('Z 13', 'Z 30,' and 'Z 30 A') evaluated at different times of the collection showed resistance to FOC race 1 GCV [01210] according to the method for the differentiation of susceptibility or resistance to *Fusarium* wilt described by [18, 19]. On the other hand, it is easy to perform the work of resistance evaluation with FC of the fungus and reduce the response time necessary to know if a plant is resistant or susceptible (48 hours). Finally, it allows the evaluation of a significant number of samples under laboratory conditions, thus accelerating genetic improvement programs for this disease. This shows the potential of biotechnology in the field of crop genetic improvement.

3. Conclusions

The creation of effective methods for the early selection of resistance to *Fusarium oxysporum* f. sp. *cubense* serves as a valuable tool for conducting precise studies on the plant-pathogen interaction and enhancing the efficiency of genetic improvement programs in this crop. Moreover, the method for distinguishing susceptibility from resistance to *Fusarium* wilt, outlined in this chapter, has the potential for application across various pathogen populations. It can prove beneficial in both conventional and biotechnological breeding programs for *Musaceae*.

DOI: http://dx.doi.org/10.5772/intechopen.1003201

Acknowledgements

This research was supported by the International Foundation for Science (Stockholm, Sweden) and by the Cuban Ministry for Science, Technology and the Environment.

Author details

Barbarita Companioni González[1*], Rómulo García Velasco[2], José Carlos Lorenzo Feijoo[3] and Ramón Santos Bermudez[4]

1 Department of Basic Sciences, Universidad Autónoma Agraria Antonio Narro, Saltillo, Coahuila, México

2 Universidad Autónoma del Estado de México, Centro Universitario Tenancingo, Tenancingo, Estado de México, México

3 Laboratory for Plant Breeding and Conservation of Genetic Resources, Bioplant Center, Universidad de Ciego de Ávila, Cuba

4 Facultad de Ciencias Agropecuarias, Universidad Técnica Luis Vargas Torres de Esmeraldas, Esmeraldas, Ecuador

*Address all correspondence to: bcompanioni2007@gmail.com

References

[1] Food and Agriculture Organization. Perspectivas a mediano plazo: perspectivas para la producción y el comercio mundial de bananos y frutas tropicales 2019-2028. 2020. Available from: http://www.fao.org/economic/est/est-commodities/banano/es/ [Accessed: Nov. 08, 2021]

[2] Ploetz RC. *Fusarium* wilt of banana. Phytopathology. 2015;**105**:1512-1521. DOI: 10.1094/PHYTO-04-15- 0101-RVW

[3] Dita M, Teixeira LAJ, O'Neill W, Pattison AB, Weinert MP, Li CY, et al. Current state of *Fusarium* wilt of banana in the subtropics. Acta Horticulturae. 2020;**1272**:45-56. DOI: 10.17660/ActaHortic.2020.1272.7

[4] Martínez GE, Rey JC, Pargas RE, Manzanilla EE. Marchitez por *Fusarium* raza tropical 4: Estado actual y presencia en el continente americano. Agronomía Mesoamericana. 2020;**31**(1):259-276. DOI: 10.15517/am.v31i1.37925

[5] Saraswathi MS, Kannan G, Uma S, Thangavelu R, Backiyarani S. Improvement of banana cv. Rasthali (silk, AAB) against *Fusarium oxysporum* f.sp. *cubense* (VCG 0124/5) through induced mutagenesis: Determination of LD50 specific to mutagen, explants, toxins and *in vitro* and *in vivo* screening for *Fusarium* wilt resistance. Indian Journal of Experimental Biology. 2016;**54**(5):345-353

[6] Buddenhagen IW. Understanding strain diversity in *Fusarium oxysporum* f. sp. *cubense* and the history of the introduction of 'tropical race 4' to better manage banana production. Acta Horticulturae. 2009;**828**:193-204. DOI: 10.17660/ActaHortic.2009.828.19

[7] Rowe P, Rosales F. Diploid breeding at FHIA and the development of Goldfinger (FHIA-01). Infomusa. 1993;**2**(2):9-11

[8] Stover RH, Buddenhagen IW. Banana breeding: Polyploidy, disease resistance and productivity. Fruits. 1986;**41**:175-191

[9] Huang BZ, Xu LB, Molina AB. Preliminary evaluation of IMTP-III varieties and local cultivars against Fusarium wilt disease in southern China. In: Molina AB, Xu LB, Roa VN, Van den Bergh I, Borromeo KH, editors. Advancing Banana and Plantain R&D in Asia and the Pacific, Proceedings of the 3rd BAPNET Steering Committee Meeting Held in Guangzhou, China, 23-26 November. Rome: IPGRI; 2004. Vol. 13. pp. 187-192

[10] Zisil X, Xin Z, Yeyuan C, Shirong L, Shouxing W. Assessment of banana germplasm for resistance to *fusarium* wilt. Chinese Journal of Tropical Crops. 2009;**3**:362-364. Available from: http://caod.oriprobe.com/articles/15644270/Assessment_of_Banana_Germplasm_for_Resistance_to_Fusarium_Wilt.htmttp://file:///C:/Users/Asus/Downloads/IN050500_eng.pdf [Accessed: Nov. 08, 2021]

[11] Molina AB, Fabregar EG, Ramillete EG, Sinohin VO, Viljoen A. Field resistance of selected banana cultivars against tropical race 4 of *Fusarium oxysporum* f. sp. *cubense* in the Philippines. Phytopathology. 2011;**101**:S122. DOI: 10.13140/RG.2.2.16107.23844

[12] Houbin C, Chunxiang X, Qirui F, Guibing H, Jianguo L, Zehuai W, et al. Screening of banana clones for resistance to *fusarium* wilt in China. In: Molina AB, Xu LB, Roa VN, Van den Bergh I, Borromeo KH, editors.

Advancing Banana and Plantain R&D in Asia and the Pacific, Proceedings of the 3rd BAPNET Steering Committee Meeting Held in Guangzhou, China, 23-26 November. Rome: IPGRI; 2004. Vol. 13. pp. 165-174

[13] Hwang SC, Ko WH. Cavendish banana cultivars resistant to *Fusarium* wilt acquired through somaclonal variation in Taiwan. Plant Disease. 2007;**88**(6):580-588. DOI: 10.1094/PDIS.2004.88.6.580

[14] Li WM, Dita M, Wub W, Hua GB, Xie JH, Geb XJ. Resistance sources to *Fusarium oxysporum* f. sp. *cubense* tropical race 4 in banana wild relatives. Plant Pathology. 2015;**64**:1061-1067. DOI: 10.1111/ppa.12340

[15] Švábová L, Lebeda A. *In vitro* selection for improved plant resistance to toxin-producing pathogens. Journal of Phytopathology. 2005;**153**(1):52-64. DOI: 10.1111/j.1439-0434.2004.00928.x

[16] Companioni B, Arzola M, Rodríguez Y, Mosqueda M, Pérez MC, Borrás O, et al. Use of *in vitro* culture-derived *Fusarium oxysporum* f. sp. *cubense* race 1 filtrates for rapid and non-destructive differentiation of field-grown banana resistant from susceptible clones. Euphytica. 2003;**130**:341-347. DOI: 10.1023/a:1023027604627

[17] García-Velasco Rómulo, Ventura-Martín José de La Caridad, Lorenzo-Feijoo José C, Portal-González Nayanci, Santos-Bermúdez Ramón, Rodríguez-García Armando et al. Evaluation of resistance to Fusarium wilt race 1 in new banana clones. Revista Mexicana de Fitopatología [online magazine]. 2023;**41**(2):285-297. DOI: 10.18781/r.mex.fit.2304-1. Epub 11 August 2023

[18] Companioni B, Mora N, Díaz L, Pérez A, Arzola M, Espinosa P, et al. Identifying discriminant factors alters the treatment of resistant and susceptible banana leaves with *Fusarium oxysporum* f. sp. *cubense* culture filtrates. Plant Breeding. 2005;**124**:79-85. DOI: 10.1111/j.1439-0523.2004.00997.x

[19] Companioni B, Lorenzo JC, Santos R. Nuevo Método Para la Diferenciación al Mal de Panamá en Banano. Spain: Editorial Académica Española; 2012. Available from: http://www.eae-publishing.com [Accessed: Nov. 05, 2022]

[20] Carlier J, De Waele D, Escalant JV. Evaluación global de la resistencia de los bananos al marchitamiento por *Fusarium*, enfermedades de las manchas foliares causadas por *Mycosphaerella* y nemátodos. Evaluación de comportamiento. In: Vecina A, Picq C, editors. Guías Técnicas INIBAP 7. Francia: Red Internacional para el Mejoramiento del Banano y el Plátano Montpellier; 2003. pp. 5-14

[21] Novak FJ, Brunner H, Afza R, Morpurgo R, Upadhyay RK, Van Duren M, et al. Improvement of *Musa* through biotechnology and mutation breeding. In: INIBAP, editor. Proceeding of the Workshop on "Biotechnology Applications for Banana and Plantain Improvement". San Jose. Costa Rica. Montpellier. France. Rome: FAO/IAEA; 1992. pp. 143-158

[22] Ribeiro LR, Amorim AP, Cordeiro ZJM, Silva S, Dita MA. Discrimination of banana genotypes for *Fusarium* wilt resistance in the greenhouse. Acta Horticulturae. 2011;**897**(52):381-386. DOI: 10.17660/ActaHortic.2011.897.52

[23] García R, Portal N, Santos R, Yanes E, Lorenzo JC, Companioni B. Método rápido aplicado en la evaluación previa de la resistencia del banano al *Fusarium oxysporum* f. sp. *cubense*.

Revista Mexicana de Fitopatología. 2020;**38**(2):389-397. DOI: 10.18781/R.MEX.FIT.2004-1

[24] Pérez L, Batlle A, Fonseca J, Montenegro V. Reaction of FHIA hybrids and landraces cultivars to *Fusarium* wilt caused by *Fusarium oxysporum* f. sp. *cubense*. In: Abstracts of the International Congress on Banana: Harnessing Research to Improve the Livelihoods. Malasia: CGIART; 6-9 July, 6-9. 2004. p. 152. Available from: http://file:///C:/Users/Asus/Downloads/IN050500_eng.pdf [Accessed: Apr. 02, 2018]

[25] Ramírez MA, Iglesias LG, Luna M, Castro AA. *In vitro* phytotoxicity of culture filtrates of *Fusarium oxysporum* f. sp. *vanillae* in *Vanilla planifolia* jacks. Scientia Horticulturae. 2015;**197**:573-578. DOI: 10.1016/j. scienta.2015.10.019

[26] Mert Z, Karakaya A. Determination of the suitable inoculum concentration for *Rhynchosporium secalis* seedlings assays. Journal of Phytopathology. 2003;**151**:699-701. DOI: 10.1046/j.0931-1785.2003.00770.x

[27] Li CY, Mostert G, Zuo CW, Beukes I, Yang QS, Sheng O, et al. Diversity and distribution of the banana wilt pathogen *Fusarium oxysporum* f. sp. *cubense* in China. Fungal Genomics and Biology. 2013;**3**:2-6. DOI: 10.4172/2165-8056.1000111

[28] Portal N, Soler A, Ribadeneira C, Solano J, Portieles R, Herrera L, et al. Phytotoxic metabolites produced by *Fusarium oxysporum* f. sp. *cubense* race 2. Frontiers in Microbiology. 2021;**12**:1-8. DOI: 10.3389/fmicb.2021.629395

[29] García R, Portal N, Santos R, Rodríguez A, Companioni B. Mejoramiento genético para la resistencia a marchitez por *Fusarium* en banano. Revista Mexicana de Fitopatología. 2021;**39**(1):122-146. DOI: 10.18781/R.MEX.FIT.2008-2

[30] Ventura JC, López J, Rodríguez S, Medero V, González L, Ventura V, et al. Obtención de mutantes de porta bajo por irradiaciones en el cultivar 'Zanzíbar' (*Musa* spp, grupo AAB). Revista Agricultura Tropical. 2015;**1**(1):39-50

[31] García R, de La Caridad VJ, Lorenzo JC, Portal N, Santos R, Rodríguez A, et al. Evaluación de la resistencia a marchitez por *Fusarium* raza 1 en nuevos clones de banano. Revista Mexicana de Fitopatología. 2023;**41**(2):285-297. DOI: 10.18781/R.MEX.FIT.2304-1

[32] Dita M, Barquero M, Heck D, Mizubuti ESG, Staver CP. *Fusarium* wilt of banana: Current knowledge on epidemiology and research needs toward sustainable disease management. Frontiers in Plant Science. 2018;**9**:1468. DOI: 10.3389/fpls.2018.01468

Section 2

Toxin Production and Controlling of *Fusarium* Pathogens

Chapter 4

Role of Plant Defence System in Crop Protection against *Fusarium* Pathogens

Nadeem Iqbal, Riyazuddin Riyazuddin, Muhammad Nauman, Zalán Czékus, Malik Tahir Hayat, Péter Poór and Attila Ördög

Abstract

Fusarium pathogens are causal agents of several crop diseases and produce harmful mycotoxins resulting in crop and yield reduction worldwide. Among crop diseases, *Fusarium* wilt, *Fusarium* head blight, and *Fusarium* root blight are mostly reported diseases in numerous vegetables, crops, and fruits and have posed pressure on current food production and safety. In addition, the production of mycotoxins further aggravates plant health and causes serious health risks in humans and animals through food chain contamination. Different management practices have been enlisted in this chapter to reduce or eradicate *Fusarium* wilt in different crops. Interestingly, various mechanisms developed by plants have also been highlighted to fight against *Fusarium* pathogens and limit the growth of mycotoxins. One of defence mechanisms is plant antioxidant mechanisms to reduce oxidative stress by increasing enzymatic and non-enzymatic antioxidants to maintain cellular homeostasis under *Fusarium* infection. The other defence response is through hormonal signalling to combat fungal pathogens. Different phytohormones such as salicylic acid, ethylene, jasmonate, abscisic acid, cytokinin, auxin, and other plant secondary metabolites play a crucial part in the reduction of *Fusarium* growth and inhibit mycotoxin production through defence-related genes. Further, the use of different pre-harvest and post-harvest strategies has been elucidated to enhance plant resistance and growth by decreasing fungal pathogenicity and virulence.

Keywords: antioxidants, crop protection, *Fusarium* pathogens, mycotoxins, phytohormones

1. Introduction

World's population is growing exponentially and will reach 10 billion by 2050, adding pressure on available food production, its demand, and safety [1]. On the contrary, different environmental factors such as climate change, diseases, pests, and droughts are severely affecting current food supply and production system [2]. Different plant diseases cause crop reduction and yield losses which are major threats to food security globally. These diseases caused by various plant pathogens such as bacteria, fungi, and viruses can pose severe plant damage leading to economic losses [3, 4]. Among plant pathogens, *Fusarium* species are considered important plant pathogens causing wilting, necrosis, plant cell death, and mycotoxin production [5]. These *Fusarium* pathogens are responsible for many plant diseases such as head blight of root and shoot, root rot, and vascular wilt of numerous plants such as barley, wheat, grasses, etc. [6, 7]. The presence of mycotoxins in stored food and various diseases caused by *Fusarium* species in agricultural crops have attracted the attention of global researchers to study more about *Fusarium* species and different strategies to control or inhibit their growth in the agriculture sector [8]. Therefore, it is necessary to take serious action to control the growth of these pathogens to protect plants from disease development and mycotoxin production.

Fusarium pathogens and their mycotoxins cause huge destruction in the agriculture sector and are responsible for different diseases, yield reduction, and economic losses. Fungal species can infect grain causing cereal fusariosis [9]. The fungal pathogens survive on the post-harvest residues as saprotrophs or some fungi as chlamydospores [10]. The members of *Fusarium* genus cause *Fusarium* blight of head and root, death of seedlings, and production of harmful toxins. Cereal crops in flowering stage are more vulnerable to pathogenic infection, especially under humid, warm, enriched dew, and rainfall season [11]. *Fusarium* species have also been reported to cause changes in storage protein and harm to starch granules in the kernels; eventually the mycotoxin production can further reduce grain quality [12]. *Fusarium* head blight is the most devastating disease and damages many cereal crops, especially wheat and oat. However, different favorable conditions promote the *Fusarium* growth and development such as high temperature (above 20°C), high humidity, and pathogen inoculum, and cereal cultivation can cause 50% reduction of grain yield [13, 14]. Hence, it is mandatory to take preventive measures in order to sustain agriculture crops and reduce the negative impacts of these *Fusarium* pathogens.

Plants are equipped with various plant defence mechanisms to combat these phytopathogens, and these mechanisms include plant physical barriers (structural changes), changes in different metabolic pathways to produce plant secondary metabolites (chemical changes), plant antioxidant system, hormonal signalling, and expression of defence-related genes [15–17]. Therefore, to prevent *Fusarium* pathogens and their infestations, several approaches have been used such as development of resistant varieties, genetically modified plants with higher resistance, and use of fungicides with their certain limitations. As plant breeding and genetic modifications required more time and sometimes pathogens become resistant in their hosts. In the case of fungicides, environmental contamination can alarmingly affect soil microflora [18, 19]. It is obvious that *Fusarium* species cause different crop diseases and produce mycotoxins which impose negative impacts on crop production and global food security. Therefore, urgent measures should be carried out to stop the growth of such pathogenic fungal species. Moreover, efficient and safe control measures are needed to be developed for sustainable agriculture and must be implemented to end hunger, increase food security, and promote sustainable agriculture.

1.1 *Fusarium* species as plant pathogens

Fusarium genus consists of both pathogenic and nonpathogenic fungal species but still taxonomically complex [20, 21]. *Fusarium* species, especially pathogenic species, are characterized as crop pathogens and mycotoxin producers. Various *Fusarium* species have been isolated from various media such as soil and plant material and are known as pathogens, endophytes, ascomycetes, and saprobes [22]. The members of this genus are identified as pathogens in crops such as potato, rice, maize, beans, tomato, banana, mangoes, wheat, and sugar cane [23, 24]. Several environmental factors including climate, humidity, temperature, geographical area, and seasons can influence the growth and development of these pathogens [25]. The most well-known pathogens of this genus are *F. oxysporum*, *F. moniliforme*, and *F. graminearum* which have severely affected many cereal crops, vegetables, and fruits [15, 26]. The diseases caused by these species include seed blight, pokkah boeng, Panama disease, head blight, bakane, vascular wilts, and rots of different plant parts such as stem, crown, root, and ear [8, 27]. Therefore, *Fusarium*-caused diseases can have negative impacts on crop production, yield, and storage after harvesting and consequently pose devastating effects on food security by reducing their quality and quantity.

These pathogenic species may enter into plant roots from soil or other media such as air, ground, and water. In addition, other factors including insects, injuries caused by new roots, and nematodes can lead to disease development or its symptoms such as chlorosis, necrosis, and wilting (**Figure 1**) [28]. The direct attack of *Fusarium* species in the field can cause economic damage which is further worsened due to the production of mycotoxins or allergenic compounds during post-harvest storage. These species produce very toxic secondary metabolites including diacetoxyscirpenol, fumonisins, zearalenone, deoxynivalenol, fusaric acid, and nivalenol which

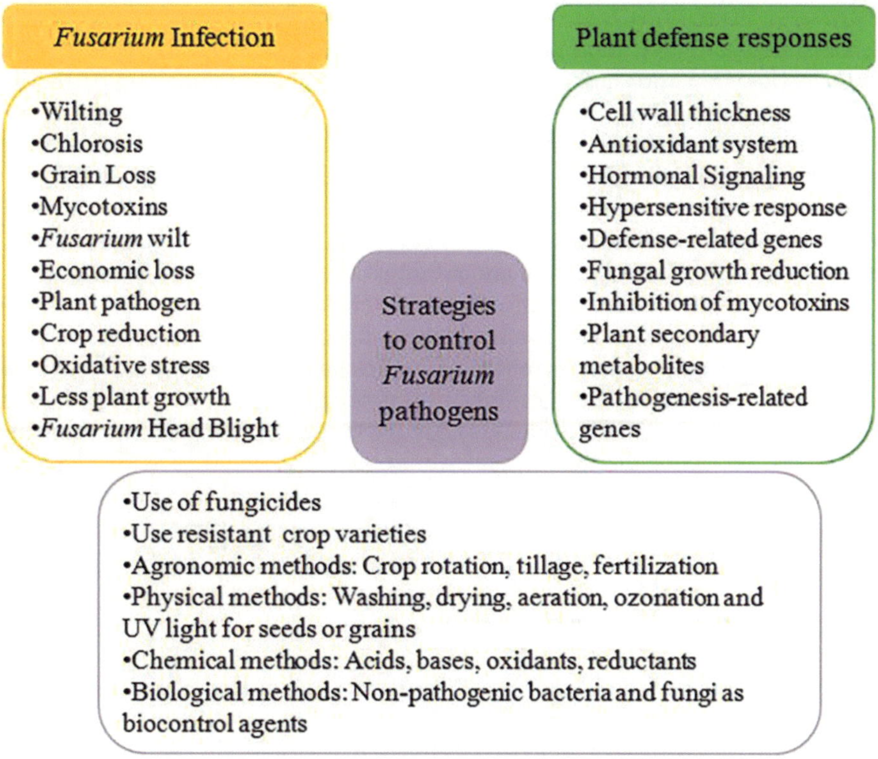

Figure 1.
Various effects caused by Fusarium *infection and different plant defence responses under* Fusarium *attack. The use of different strategies to control* Fusarium *pathogens at pre- and post-harvest stages can reduce pathogenic infection and mycotoxin production.*

have deleterious effects on humans and animals due to contaminated food and feed, respectively [29, 30]. The food contamination due to mycotoxins is a global concern in several countries because their presence has affected around 25% of the world's crops [31]. Further, the contamination of food products is also linked with different health risks, i.e. carcinogenesis, neurogenesis, and mutagenesis [32]. Hence, based on the toxic effects on *Fusarium* species and mycotoxin contamination in food and feed, more accurate and proper precautions should be taken while crop cultivation and food storage to limit their growth and reduce the occurrence of mycotoxins.

2. Pathogen perception and plant defence system

Plants have evolved specific and effective defence systems to recognize *Fusarium* pathogens. Plants perceive several signalling molecules both from *Fusarium* species and from their own cell walls upon *Fusarium* contact [33]. However, different plants exhibit different mechanisms of defence against these *Fusarium* attacks. The first plant barrier is the plant cell wall which *Fusarium* pathogens encounter, and this barrier determines the strength of the host plant. For instance, some banana plants showed resistance against *F. oxysporum* Foc TR4 even before fungal colonization [34]. The resistance in these banana varieties is due to the presence of different genes such as polyphenol oxidase, cellulose synthase, and glutathione S-transferase which provide strength to the cell wall during pathogen attack in resistant banana plants [34]. Plants possess different pattern recognition receptors (PRRs) to detect the presence of microbial-associated molecular patterns (MAMPs) or pathogen-associated molecular patterns (PAMPs) and send downstream signals to initiate signalling pathways leading to plant immunity [35]. Hence, the plant defence system is triggered by PRRs followed by the transcription of pathogenesis-related (*PR*) genes. The production of reactive oxygen species (ROS) and other metabolites are also activated [27]. In addition, the production of endogenous and exogenous elicitors from the cell wall and cuticle provides the basal defence against *Fusarium* attacks. On the other side, *Fusarium* species also destroy pectin and secrete different fragments consisting of galacturonic acid which serves as signalling molecules and strengthen plant defence systems [36]. In addition, the main components of signalling pathways are chitin elicitor-binding protein and receptor kinase and their coding genes that recognize chitin oligosaccharides. Further, several defence-related genes such as *PR1*, *PR5*, thionins (*Thi2.1*), and plant defensin (*pdf1.2*) are induced in plants upon *Fusarium* infection [37]. So, plants induce various cellular, metabolic, and growth changes upon recognition of *Fusarium* pathogens and initiate their defence responses through defence-related genes and ROS-induced hypersensitive responses.

3. Host perception and *Fusarium* virulence

Fusarium species such as *F. oxysporum* require host plants for infection and virulence. These pathogens recognize a particular host plant and penetrate into its vascular tissues [38]. Interestingly, pathogens avoid the plant defence system during attack or infection. The presence of hydroxyl acids such as dihydroy-C16 and trihydroxy-C18 provides the safe intact of the pathogen to the host due to their unique characteristic [39]. During the infection process, *Fusarium* species perceive various physio-chemical signals from the host and show responses through various

DOI: http://dx.doi.org/10.5772/intechopen.1004924

morphogenetic and metabolic changes to enhance *Fusarium* development [40]. Moreover, hydrolases of the host also act as signals to synthesize pectinases, cutinases, and other hydrolytic enzymes essential for the penetration in the host plant [41]. During the fungal infection, various signalling cascades are triggered in fungal pathogens which are mainly dependent on the perception of signalling protein receptors from the host. For instance, Msb2 and Sho1 proteins are involved in the infection process of *F. oxysporum* [42]. These proteins contribute to *Fusarium* virulence by up-regulating different Fmk1-modulated genes encoding various enzymes involved in cell wall synthesis, taking part in cell all responses, and assisting in Fmk1-based virulence. Both these proteins are also responsible for the up-regulation of cell all biogenesis genes such as *gas1*, *chsV*, and *fks1* [39]. Thus, *Fusarium* species also deploy various defence-related actions for virulence and disease development by competing against the host defence system.

4. Production of *Fusarium* mycotoxins during plant-pathogen interaction

Fusarium species produce different toxic chemicals during plant-pathogen interaction that are harmful to humans, animals, and plants. Controlling *Fusarium* pathogens is hard because of their genetic variability and wide range of hosts [43]. *Fusarium* species are found as biotrophic, hemibiotrophic, and necrotrophic. Upon host contact, *Fusarium* acts as biotrophic to colonize host plant and then shifts to necrotrophic to produce mycotoxins and other cellulolytic enzymes to take control of the host plant [29]. Different cereal crops such as oats, wheat, maize, and barley are affected due to the production of mycotoxins by pathogenic *Fusarium* species. Further, these pathogens and their mycotoxins account for 50% yield loss in tropical fruit crops including pea, tomato, banana, lentils, and pineapple [44]. The presence of *Fusarium* toxins such as trichothecenes, fumonisins, zearalenone, fusarins, fusaric acid, moniliformin, enniatins, and beauvericins is also considered to cause infection symptoms in plants [29]. Trichothecenes are considered best toxins due to their toxicity to living organisms and are biosynthesized after 7 to 10 enzymatic modifications. The *TRI* gene cluster is responsible for the biosynthesis [45]. These mycotoxins interfere with protein synthesis by blocking elongation of polypeptide chains, intervening peptidyl transferase, and hindering chain termination [46]. Trichothecenes are further classified on the basis of polysomal breakdown as type I and type II. Type I contains T-2 toxin, fusarenon X, verrucarin A, and nivalenol, while type II trichothecenes are DON, verrucarol, and crotocin [29]. Therefore, it is essential to take serious action against mycotoxin production in several crops and somehow control mycotoxin-producing genes in the pathogens.

Fumonisins are polyketide-derived mycotoxins produced by various pathogens such as *F. verticillioides*, *F. sacchari*, *F. proliferatum*, and *F. subglutinans*. The *FUM* gene cluster is involved in the biosynthesis of these mycotoxins [47, 48]. These mycotoxins are also phytotoxic; however, fumonisin B1 has been reported widely to damage crops and other cereals. FB1 contamination results in the degradation of endosperm, disruption of plasma membrane, accumulation of sphingolipid intermediates by inhibiting ceramide synthase enzyme [15]. Maize seeds soaked with fumonisins exhibited around 75% reduction of radical elongation as compared to controls [49]. Zearalenone is involved in high electrolytic leakage, blocking of H+ release resulting in acidification and causing root length in maize and red beet. In addition, this toxin can change the permeability of both tonoplast and plasmalemma [50]. Fusarins are also polyketide-derived compounds produced by *F. graminearum*, *F. culmorum*, *F. avenaceum*, *F. fujikuroi*, *F. poae*, and *F. oxysporum* [51]. Among different types of fusarins A, B, C, and D, fusarin C has been identified as mutagen due to the presence of the C13-14 epoxide ring, while A and D are not mutagens due to lack of this ring [52].

Fusaric acid can also cause phytotoxic effects and result in wilt symptoms such as tomato wilt caused by *F. oxysporum*. This toxin is not involved in the infection phase but promotes pathogenesis in the second phase (necrotrophic stage) [53]. Fusaric acid can significantly enhance ROS production, reduce the activities of antioxidants, and resultantly lead to plant cell death [54, 55]. Exogenous application of this toxin results in the disruption of cell structure, reduction in photosynthetic rate, and decreased cellular metabolism [56]. Moniliformin is another *Fusarium* toxin produced by *F. proliferatum, F. avenaceum, F. oxysporum,* and *F. subglutinans*. It has been reported to reduce photosynthetic pigments, leaf development, and plant biomass [57]. Similarly, other *Fusarium* toxins are beauvericins and enniatins which are commonly found in food and feeds [58]. Beauvericins can cause apoptosis and DNA fragmentation due to the disruption of the mitochondrial pathway [59]. There are several environmental factors which affect the synthesis of these *Fusarium* toxins. These factors include temperature and moisture content, pH, nitrogen source, and plant extracts which determine the growth, infection, development, and mycotoxin production [29]. The production of these mycotoxins is a next step for *Fusarium* pathogens after their infection or disease symptoms. Most of these mycotoxins are produced during infection or post-harvest storage to contaminate food. For that reason, it is crucial to make progress in this regard to reduce their production and associated toxic effects.

5. How plants deal with *Fusarium* mycotoxins

Plants employ mechanisms such as chemical modification and compartmentation of *Fusarium* mycotoxins. UDP glycosyltransferase (UGT) enzyme is involved in the detoxification of deoxynivalenol mycotoxin in *Arabidopsis* [60]. Similarly, around 6 transcripts showed an effective role for trichothecene detoxification in barley [61]. Many UGTs not only perform function in the detoxification process but also act on secondary metabolites including terpenes, hormones (salicylic acid, cytokinin, auxin), and flavonoids which play regulatory roles in plant defence responses against fungal diseases [62]. Besides, plants also deploy transportation of *Fusarium* mycotoxins using multidrug transporters and their roles were confirmed by transcriptomic studies. These transporters include ATP-binding cassette (ABC), resistance nodulation-cell division family (RND), multidrug resistance family (MDR), major facilitator, and multidrug and toxic compound extrusion family (MATE) which are responsible for binding with the toxic compounds and transport them outside the cell [29]. The ABC transporters are present in tonoplast, chloroplast, mitochondria, plasma membrane, and peroxisomes performing various functions including transportation of mycotoxins which is carried out by CytP40s oxidation by conjugating glucoses or glucuronide (hydrophilic compounds). Thereafter, these less toxic conjugated mycotoxins are brought to the central vacuole and transported outside by the ABC system [63]. In the case of the MATE family, toxic chemicals such as heavy metals or lethal organic compounds are excreted along with other secondary metabolites, hormones, and acids [64]. Further, the microbial transformation or detoxification occurs in the soil by releasing different enzymes, and therefore, such mycotoxins are not accumulated in the soil. Similarly, zearalenone is transformed into its reduced phase such as α-zearalenol and β-zearalenol [65]. However, more research work is required in the identification and screening of different enzymes involved in the detoxification of mycotoxins.

In addition, genetic modification either by plant breeding or transgenesis can significantly prevent pathogens and their mycotoxins. For instance, rice plants expressing the *TRI101* gene exhibited acetylase activity of deoxynivalenol in rice plants [66]. Likewise, transgenic wheat plants expressing UDP HvUGT13248 of barley resulted in the suppression of diseases in wheat spikes caused by *F. graminearum* and also showed conjugation of deoxynivalenol toxin to deoxynivalenol-3-glucoside [67]. Intriguingly, another research showed the relation between *Triticum aestivum Fusarium* resistance orphan gene (*TaFROG*) and deoxynivalenol mycotoxins and displayed that *TaFROG* enhanced wheat tolerance to the mycotoxins [68]. However, the plant-pathogen interaction and possible plant defence can determine the quality of efficacy of both plant breeding and transgenesis approaches to overcome these *Fusarium* pathogens and their harmful toxins. Consequently, the use of chemical modification, compartmentation, transporters' function, and use of transgenic plants with detoxifying genes can significantly lower the chances of *Fusarium* infection and toxin production.

6. *Fusarium* wilt and its management

Fusarium species have been reported to cause different plant diseases including chlorosis, premature leaf falling, wilting, browning of vascular system, and necrosis leading to huge reduction in crop yield [69, 70]. Due to a broad range of host specificity, *Fusarium* species are classified into numerous races and formae specials. Among them, wilt pathogens are *F. oxysporum*, *F. verticillioides*, *F. graminearum*, and *F. solani* [71]. Based on economic damages, *F. oxysporum* is the most devastating pathogen which has been documented in more than 150 host plants such as cabbage, tomato, cotton, banana, watermelon, flax, onion, pea, tulip, etc. [72]. *F. oxysporum* causes virulence in plants due to the releases of small cysteine-rich proteins [73]. It has been reported that *F. oxysporum* commences infection through roots and then penetrates into xylem vessels leading to wilting upon colonization and restricting water movement in vascular bundles [74]. Further, pathogenesis and invasion into host plants is carried out by pathogen-produced toxic metabolites [75]. Similarly, *F. graminearum* causes crown rot and *Fusarium* head blight of cereal crops causing yield losses annually [76]. *Fusarium* head blight is a well-known fungal disease worldwide leading to economic losses of important crops such as maize, barley, and wheat [77]. This pathogen causes compromised quality of seed due to infestation and food chain contamination due to the presence of mycotoxins affecting human and animal health [78]. As the *Fusarium* wilt is the major cause of crop reduction reported by various studies, therefore, its management is of significance to lessen the effects of this disease in agriculture.

Fusarium wilt is a major problem globally, and several control strategies have been practiced for these fungal pathogens. The success rate of controlling such pathogens is limited due to the discoveries of new pathogens or their races [79]. Various methods used to control *Fusarium* wilt include resistant development, biological, cultural, and chemical (fungicides or natural products) [80]. The effectiveness of different fungicides such as thiophanate-methyl, acibenzolar-S-methyl, and prothioconazole have been evaluated and found efficient in reducing wilt symptoms in watermelons grown in the field [81]. Other possible options to reduce fungal growth are soil fumigation using different chemicals such as propylene oxide, methyl isothiocyanate, chloropicrin, and sodium azide. Further, cover cropping, use of agrochemicals, and crop rotation are also employed to avoid fungal infection [76]. Furthermore, different plant growth-promoting bacteria (PGPB) are also used as a biocontrol against these pathogens. For instance, species of *Pseudomonas* and actinobacteria have shown effective control of *Fusarium* wilt [82, 83]. Moreover, the use of mixing bacterial cultures (*Bacillus subtilis* OH 131.1 and *Cryptococcus flavescens*) exhibited efficient control of *Fusarium* head blight in wheat [84]. Similarly, another research using the combination of three bacterial strains such as *B. amyloliquefaciens*, *Paenibacillus polymyxa*, and *B. subtilis* reported the highest efficacy rate against *F. graminearum* in wheat [85]. These results show that combinations of different biological control agents can be used for effective control against *Fusarium* pathogens.

7. Plant antioxidant system to combat fungal pathogens

Fusarium pathogens induce ROS production in plants which affect various plant processes including growth, development, and other metabolic processes leading to reduced crop production and yield [86]. In order to cope with ROS accumulation under *Fusarium* attacks, plants have developed vigorous antioxidant systems to detoxify ROS accumulation and promote plant growth [87]. This antioxidant system is composed of enzymatic and non-enzymatic antioxidants to protect plants from harmful effects of ROS overproduction such as superoxide ($O_2^{\cdot-}$), hydrogen peroxide (H_2O_2), hydroxyl radical ($^{\cdot}OH$), and singlet oxygen ($^1O_2$) [15]. These antioxidants include superoxide dismutase (SOD), catalase (CAT), peroxidase (POD), polyphenol oxidase (PPO), ascorbate peroxidase (APX), ascorbate (ASA), and glutathione (GSH) and perform scavenging function of ROS production during pathogenic attacks [17, 26].

Fusarium infection was reported in different parts of banana which significantly enhanced production of SOD, POD, APX, PPO, and lipoxygenase under oxidative stress and prevented plant cell death [88]. On the other hand, ROS production is considered essential for hypersensitive response to perform its function properly [88]. Further, tomato plants treated with SA showed higher reduction in plant root and shoot dry weight and exhibited higher accumulation of H_2O_2 and lipid peroxidation under *F. oxysporum* attack. The antioxidant activities of SOD and glutathione peroxidase (GPX), and ASA were increased in infected plants when treated with SA, MeJ, and *Trichoderma harzianum* which showed resistance against *Fusarium* wilt [89]. Furthermore, *F. oxysporum* induced *Fusarium* wilt in melons and caused drastic yield loss. The effect of this pathogen was studied in two resistant (Shante-F1 and Khatooni) and two susceptible (Shante-T and Shahabadi) melon cultivars and found that *Fusarium* infection significantly induced production of SOD, CAT, POX, PAL, PPO, chitinase (CHI), β-1,3-glucanase (GLU), and total phenolics in the roots of resistant varieties [17]. However, no significant difference was observed in fresh weight, dry weight, diameters, height, and biomass among resistant and susceptible cultivars [17]. Similarly, tomato plants also exhibited higher activities of SOD and APX antioxidants under *F. oxysporum* infection, while glutathione reductase (GR) and dehydroascorbate reductase (DHAR) enzymatic activities were also enhanced when plants were inoculated with arbuscular mycorrhizal fungi such as consortium of *Claroideoglomus etunicatum*, *Funneliformis mosseae*, and *Rhizophagus intraradices* [90]. Hence, activation of antioxidant enzymes under fungal attack clearly suggests the regulator role of these antioxidants against *Fusarium* pathogens.

Guaiacol is a natural antioxidant and antifungal agent and has been studied against *F. graminearum* in wheat which inhibited mycelia growth, germination and formation of conidia, and production of mycotoxin deoxynivalenol [91]. Guaiacol (1.838 mM) reduced levels of malondialdehyde (MDA; indicator of lipid peroxidation) and activities of SOD, POD, and CAT ith increasing concentration of guaiacol from 1.6 to 6.4 mM [92]. In addition, *Fusarium* wilt caused in cucumbers due to *F. oxysporum* was reduced with silicon treatment and enhanced plant resistance by improving POD, CAT, and APX activities and reducing MDA levels [92]. Moreover, other parameters such as transpiration rate, stomatal conductance, net photosynthetic rate, RuBisco activity were also elevated under silicon treatment on infected plants indicating the important role of silicon against *Fusarium* wilt [92]. The effect of nitrogen treatments was studied in nine varieties of wheat against *F. culmorum*, and the activities of different antioxidant enzymes such as APX, CAT, and GR were measured. The low nitrogen levels enhanced antioxidant activities of some enzyme under *Fusarium* infection while high nitrogen levels significantly reduced antioxidant activities [93]. Additionally, the plant antioxidant system was also activated when three different cyanobacteria such as *Anabaena oryzae*, *Desmonostoc muscorum,* and *Arthrospira platensis* were inoculated with pepper plants (*Capsicum annuum* L.) infected with *F. oxysporum* [94]. The results showed that addition of cyanobacteria significantly reduced *Fusarium* infection, enhanced POD and PPO expressions, and also increased SA, indole-3-Acetic levels, while decreased ABA content in infected plants showing that use of foliar treatment of these bacteria can improve plant growth, metabolic characteristics, and defence responses against *Fusarium* wilt [94]. Likewise, the activities of SOD, POD, and CAT antioxidants were documented higher in *Medicago truncatula* plants when treated with *Sinorhizobium meliloti* and *Sinorhizobium medicae*, and these bacterial treatments also increased sucrose content and reduced oxidative stress due to fungal pathogen *F. oxysporum* [95]. Thus, it is confirmed from various reported studies that the plant antioxidant system is activated under *Fusarium* attack and plays its vital part by detoxifying excess ROS production and improving plant growth and development. Further, the use of different exogenous antioxidants or symbiotic bacteria and fungi can improve plant defence responses with higher activities or levels of antioxidants.

8. Role of secondary metabolites in plant resistance

Various plant secondary metabolites are produced upon pathogenic attack or mycotoxin production. These secondary metabolites are involved not only in killing *Fusarium* pathogens but also in inhibiting mycotoxin production [86]. For instance, terpenoids and phenylpropanoids have been reported to provide a basic defence line for pathogenic fungi [96]. Other studies also documented the role of ferulic acid, chlorogenic acid along with small amounts of carotenoids, α-tocopherols, and putrescine in *Fusarium*-infected kernels of maize plants [86, 97]. Further, the production of mycotoxins such as HT-2 and T-2 by *F. sporotrichioides* and *F. langsethiae* was significantly decreased upon phenolic acid treatments which also act as plant antioxidant to reduce oxidative stress on pathogenic invasion [98]. Furthermore, plants synthesize ferulic acid upon pathogen recognition which also inhibits the transcription of various trichothecene genes such as *TRI5*, *TRI6*, and *TRI12*. The fumonisin production was also blocked by α-tocopherols; however, carotenoids did not show an effect on the production of mycotoxins [29]. Interestingly, trichothecene-producing *F. graminearum* strain infected barley and induced production of different secondary metabolites such as cinnamic acid, naringin, sinapyl alcohol, geranyl chalconaringenin, heptadecatrienoic acid, dihydroxy linoleic acid, and dihydroquercetin which ultimately enhanced plant resistance and also produced JA phytohormone [99]. Similarly, graminaceous plants also produce benzoxazinoids in the shikimate pathway and other 20 metabolites which have exhibited resistance against fungal pathogens [100]. It is obvious from different scientific reports that plant secondary metabolites can also enhance plant defence systems by reducing fungal growth and inhibiting toxin-producing genes in the pathogens.

9. Phytohormonal Defence Signalling against *Fusarium* Pathogens

Plant hormones play a crucial role in defence mechanisms to protect plants from *Fusarium* pathogens and their mycotoxins. An interconnected network of complex signalling is stimulated downstream of effector-triggered immunity (ETI) and pathogen-triggered immunity (PTI) against pathogens [101]. Further, PTI initiates different cellular responses including ROS production, changes in the ionic flux of the cytosol, induction of mitogen-activated protein kinase, hormonal defence, and other physical barriers such as cell wall thickness [102]. The main plant hormones involved in plant defence signalling against pathogens are abscisic acid (ABA), ethylene (ET), jasmonate (JA), and salicylic acid (SA) playing their defensive roles in several ways depending upon plant type, hormone concentration, and infection stage [103, 104]. Based on plant growth stage, these phytohormones can function in synergistic or antagonistic ways [105]. For instance, SA is responsible for defence against biotrophic pathogens by inducing local and systemic resistance, while ET and JA have been found potent against necrotrophic pathogens. Hence, all these plant hormones are involved in plant defence responses by triggering induced systemic resistance (ISR). Further, other plant hormones such as auxin, cytokinins, brassinosteroids, and melatonin are also implicated in plant growth, development, and defence responses [106]. Therefore, the hormonal crosstalk protects plants from pathogens by employing different defence strategies including signalling networks [107].

SA production is linked with *PR* expressions in dicot plants such as *Arabidopsis* and tobacco, while the involvement of SA in monocots has not been fully explained under pathogenic attacks [108]. *F. graminearum*-infected wheat spikelets exhibited the role of SA with higher expressions of *PR1* and *PR4*, while *Pdf1.2*, *PR1*, and *PR4* were highly expressed in the case of *Arabidopsis* infected with *F. graminearum* elucidating the specific roles of both SA and JA [109]. Methyl salicylate (an SA form) has been found effective in signal transduction of systemic acquired response (SAR) against different pathogens such as bacteria, fungi, and viruses [110]. Further SA was also reported in wheat against *F. graminearum*, a causal agent of *Fusarium* head blight [108]. SA foliar application also showed resistance against *Fusarium* wilt and increased peroxidases and phenylalanine ammonia-lyase activities in beans [111]. Conversely, fungal pathogens avoid SA formation in order to escape from SA-mediated defence response. For instance, a causal agent of corn smut *Ustilago maydis* releases chorismate mutase (Cmu1) to interfere with SA biosynthesis. The deletion mutants of Cmu1 showed higher SA accumulation exhibiting the role of Cmu1 as an antagonist of SA formation [112]. Similarly, some fungal pathogens are also capable of producing salicylate hydroxylases to degrade SA for rapid fungal growth and infection [113]. The role of SA treatment was also documented in tomato plants against *Fusarium* wilt caused by *F. oxysporum* and showed crosstalk with nitric oxide to induce plant immunity by reducing disease incidence [114]. Likewise, *F. oxysporum* also significantly enhanced SA and terpene production in the roots and leaves of *Chrysanthemum morifolium* as compared to control plants depicting the involvement of SA in plant defence responses [115]. Hence, these findings suggest that SA either exogenously or endogenously plays an essential role in the plant defence system against *Fusarium* pathogens, and SA could be used in agriculture to prevent yield losses and enhance crop production limiting fungal pathogens.

JA also plays an important part in signal transduction to induce plant defence responses under biotic stress conditions and is also involved in SAR during plant-pathogen interactions. For instance, JA treatment in rice significantly increased plant resistance against necrotrophic pathogens by activating the phenylpropanoid pathway [116]. The important roles of JA, SA, and ET were studied in tomato plants using gene markers of the hormones against *F. oxysporum* infection and the results showed that pathogenic infection enhanced gene expressions of both SA and ET indicating their activated defence pathway [117]. *NahG* mutants exhibited higher susceptibility to *Fusarium* infection as in these mutants SA is degraded elucidating the central role of SA against pathogens. However, in the case of ET, *Never ripe* (no ET perception) and *ACD* (no ET biosynthesis) mutants reduced disease symptoms and fungal growth [117]. These findings show the positive regulation of ET and negative modulation of SA. On the contrary, *def1* (JA is compromised) and prosystemin overexpressed (JA is activated) mutants showed no changes under *Fusarium* attack [117]. Another study also reported the significance of JA and ABA in wheat plants against *F. graminearum*, and the results displayed that JA significantly reduced fungal growth and *Fusarium* head blight while ABA facilitated wheat head blight [118]. Similarly, the involvement of JA and ABA was also detected in two rice cultivars (Selenio (resistant) and Dorella (susceptible)) infected with *F. fujikuroi* (caused bakanae disease) and the outcome showed higher levels of phytoalexins in Selenio cultivar and bakanae disease symptoms were observed while susceptible cultivar (Dorella) exhibited less phytoalexin production and increased ABA and gibberellin while inhibiting JA production [119]. These studies indicate how crosstalk among plant hormones is involved in the induction of plant defence responses against fungal pathogens.

ET is a light gas molecule and is involved in the regulation of PTI to prevent fungal pathogens [120]. The ETR receptor (ETR1) is required by *F. oxysporum* for disease development. For instance, *Arabidopsis etr1* mutants showed less symptoms of *Fusarium* wilt than control plants and it also increased expressions of SA-encoding genes [121]. ET signalling is responsible for plant defence responses against fungal pathogens, for instance, ET signalling impaired *Arabidopsis* mutants showed higher resistance for *F. graminearum*; however, overexpressed mutants with ET signalling elicited *Fusarium* susceptibility confirming the integral role of ET in *Fusarium*-plant interactions [122, 123]. Similarly, the etr1-1 mutant enhanced plant resistance against *F. oxysporum* in *Arabidopsis*, suggesting that ETR1 is also needed for disease development [121]. The use of *F. oxysporum* isolate was reported for pathogenic virulence and showed that the expressions of different defence-related genes are dependent on fungal virulence [124]. Further, [125] demonstrated that tobacco plants infected with *F. solani* increased SA accumulation during initial infection and then enhanced JA production elucidating the involvement of both SA and JA for *Fusarium* resistance. Another study highlighted the role of SA and JA signalling pathways in *Arabidopsis* infected with *F. sporotrichioides* and showed higher expression of SA-responsive gene *PR1* at early stage 24 to 48 h while JA-responsive gene *PDF1.2* was highly expressed [126]. The role of NPR1 was assessed in *Arabidopsis* against *F. asiaticum* and was found responsible for disease susceptibility while showing resistance in floret assays explaining the dual role of NPR in the SA signalling pathway against *Fusarium* infection [127]. Moreover, the effects of different hormones were also studied in floral and root tissues of *Brachypodium distachyon* against *F. graminearum* infection, a causal agent of *Fusarium* root and head blight and found that SA remarkably enhanced resistance to *Fusarium* head blight while showed susceptibility to *Fusarium* root blight [128]. Conversely, JA and ET increased resistance against *Fusarium* root blight while exhibiting susceptibility to *Fusarium* head blight. Interestingly, cytokinin and Auxin showed susceptibility and resistance, respectively, against *Fusarium* root and head blight [128]. Hence, it is evident that different plant hormones play their specific roles during plant-pathogen interactions to enhance plant defence systems and prevent pathogen attacks.

10. Different strategies to control *Fusarium* pathogens

Plants employ different strategies to control *Fusarium* pathogens to promote plant growth and development. These plant strategies are pre-and post-harvest approaches including various physical, chemical, and biological methods. The use of natural products or plant extracts has also been described to limit fungal pathogens.

10.1 Biocontrol using bacterial species

Different species of *Streptomyces*, *Pseudomonas*, and *Bacillus* have been reported as biocontrol agents [129]. Nevertheless, *Bacillus* species have shown more advantages over other species due to the presence of certain characteristics such as endosperm formation, ubiquity, different pH tolerance, osmotic conditions, temperature, and lack of disease development [71, 130]. *Bacillus* species develop biofilms with plant roots and enhance plant growth by more nutrient uptake and also induce plant resistance through disintegration of fungal mycelia [131]. The production of siderophores by bacterial species can also play an important role in plant growth such as providing iron to plants and limiting it to fungal pathogens. For instance, *B. pumili* showed an efficient biocontrol activity against tomato *Fusarium* wilt through siderophore production [132]. In addition, biocontrol bacteria can secrete certain extracellular enzymes for the breakdown of the cell wall of fungal pathogens [133]. Further, *B. subtilis* SG6 showed antagonistic interaction with *F. graminearum* and hindered sporulation of fungal pathogens by lysis of fungal cell wall [134]. Similarly, another research showed an effective control of *B. thuringiensis* NM101-19 in soybeans against *Fusarium* infection [135]. Likewise, *Fusarium* wilt disease was suppressed in tomato plants due to the production of phenyl acetic acid by *B. fortis* IAGS162, and this chemical also assisted in the rhizosphere colonization of this bacterium [136]. Hence, it is clear from the reported literature that the use of different bacteria can promote plant growth by reducing pathogenic attacks through various mechanisms.

10.2 Biocontrol using fungal species

Non-pathogenic *Fusarium* species have been reported in the suppression of *Fusarium* wilt or other diseases in various crops, fruits, and cereals and vegetables by employing different mechanisms [137]. For instance, nonpathogenic *F. oxysporum* showed resistance against *F. oxysporum* f. sp. Cubens in bananas acting as plant growth promotion (PGP) [138]. Similarly, *Fusarium* species exhibited resistance in response to *F. culmorum* in flax plant through induced systemic response (ISR) [139]. Likewise, *F. oxysporum* competed for nutrition source and used ISR mechanism to enhance resistance in date palm against *F. oxysporum* f. sp. *Albedinis* [140]. Moreover, *F. oxysporum isolate (F.o-T5)* was found effective in tomato plants against *F. solani* infection using antagonistic strategy to enhance resistance in plants and promote plant growth [141]. The use of other fungal species such as yeast species have also been assessed and found efficient in controlling *Fusarium* pathogens. For example, *Cryptococcus carnescens* (E22) displayed high resistance against *Fusarium* pathogens such as *F. culmorum*, *F. poae*, and *F. graminearum* by antagonistic mechanisms and inhibited their growth by competing for space and nutrients, siderophore production, and secretion of chitinase and β-1,3-glucanase enzymes for the breakdown of fungal cell wall [142]. Thus, non-pathogenic fungal species can also be used as an effective strategy to reduce *Fusarium* pathogens by using different mechanisms such as ISR, PGP, and antagonistic approach.

10.3 Use of natural products from medicinal plants

Plant products exhibit antifungal properties and are eco-friendly in nature as these compounds do not persist in the environment for a longer period than other synthetic fungicides [143]. These plant extracts are widely being used in the agriculture sector to prevent fungal pathogens and control plant diseases. The antimicrobial property is based on the presence of certain secondary metabolites such as flavonoids, phenols, glycosides, alkaloids, polyphenols, tannins, and other compounds [144]. Numerous studies have elucidated the roles of plant products to prevent fungal infection of various *Fusarium* species such as *F. proliferatum*, *F. verticillioides*, *F. solani*, and *F. oxysporum* in vegetables, fruits, and cereal crops [145]. The leaves of *Combretum molle* R. Br have been used to control various *Fusarium* pathogens such as *F. oxysporum*, *F. verticillioides*, *F. solani*, *F. graminearum*, *F. proliferatum*, *F. equiseti*, *F. semitectum*, *F. chlamydosporum*, and *F. subglutinans* [146–148]. Similarly, antifungal compounds isolated from *Artemisia annua* L. have extensively been studied and used against *F. solani* and *F. oxysporum* [149]. Moreover, various antifungal compounds such as artemetin, dehydrodiconiferyl alcohol, denudatin A, B, Futokadsurin B, C, gallic acid, maslinic acid, penduletin, withaferin A, etc. extracted from different plant parts have shown effective control against many *Fusarium* pathogens [145]. Subsequently, using natural products or extracts can have a huge impact on fungal reduction by promoting plant growth and resistance.

10.4 Pre- and post-harvest strategies to control *Fusarium* pathogens

The purpose of these preventive measures is to reduce *Fusarium* pathogens and their production of mycotoxins. These strategies include agronomic methods such as crop rotation, tillage and fertilization, use of quality seeds, appropriate sowing dates, and weather conditions, resistant crop varieties, and biological control of plant protection [14]. In addition, the use of different chemicals such as fungicides including metaconazole, triazole, and tebuconazole can significantly decrease the chances of pathogen occurrence and their secondary metabolites, i.e. mycotoxins [150]. These pre-harvest methods have shown the impact on the reduction of pathogenic fungi and also limited inhibited mycotoxin production. Similarly, the protection of crops after harvesting is also very important, and certain actions should be taken to prevent fungal diseases and grain contamination by mycotoxins. These methods include physical, biological, and chemical (**Figure 1**). For instance, physical methods are drying, adequate aeration, sorting, grinding, washing, cleaning grain surface, suitable humidity, adequate temperature for seed storage, use of UV light illumination and adsorbents, and ozonation that have shown efficient reduction on the growth of *Fusarium* pathogens [14]. Biological methods consist of using microorganisms such as non-pathogenic bacteria and fungi which function as antagonistic to *Fusarium* pathogens and reduce their growth while limiting mycotoxin production (Sections 11.1 & 11.2). In addition, these microbes compete for nutrition sources and space, as well as trigger ISR against *Fusarium* species [151]. Some microbes are also capable of degrading or transforming these harmful toxins into less toxic or harmless compounds [152]. Moreover, the chemical methods consist of conversion of toxic chemicals into less harmful chemicals via different chemical reactions. The chemicals used in this method are different bases, acids, oxidants, and reductants. In addition, other chemicals such as ammonification (ammonium hydroxide), use of hydrogen sulfate, and 1% sodium hypochlorite have exhibited remarkable reduction in fungal diseases to protect various agricultural crops [153]. Therefore, before and after harvesting, the use of these strategies in agriculture can significantly increase crop production and yield by reducing contamination chances of cereal grains, fungal infection, and mycotoxin production.

11. Future challenges

Numerous crops are susceptible to *Fusarium* pathogens resulting in crop and yield reduction globally. The present disease management approaches become limited when plant pathogens acquire resistance against chemicals such as fungicides. In addition, the availability of resistant cultivar against these *Fusarium* pathogens is also confined which further pose pressure on food security. The identification, selection, and time required for attributes of host plants, fungal pathogens, and development of resistant plant varieties put pressure on available resources. During this process, more resistant fungal pathogens emerge and destroy crops through infection or diseases. The emergence of pathogens is faster than the discovery of new resistance cultivars or other antifungal agents. Conversely, the use of fungicides also causes environmental pollution and disturbs soil micro fauna and flora affecting soil characteristics. Further small RNAs of plants hinder expressions of fungal genes acting as an innate plant defence using an RNAi-dependent mechanism which has shown potential to reduce fungal disease by decreasing fungal virulence and pathogenicity. However, gene silencing of *Fusarium* species can be an effective approach to control *Fusarium* pathogens by avoiding chemical costs and hazards associated with them.

12. Conclusion and future perspectives

It is concluded that *Fusarium* pathogens can drastically reduce crop yield and cause various crop diseases during pre- and post-harvest stages. Plants and *Fusarium* pathogens use various defence responses against each other for pathogen resistance and host virulence, respectively. It is evident from the literature that plants are equipped with different defence systems such as plant antioxidant system, hormonal signalling, expression of defence-related genes, and production of secondary metabolites which can significantly enhance plant defence responses against fungal pathogens by reducing their growth and mycotoxin production. Further *Fusarium* wilt is the main devastating disease and can be mitigated using different management approaches. Furthermore, various physical, chemical, and biological methods should be adopted to limit fungal growth and enhance crop yield. A deep and thorough knowledge about *Fusarium*-plant interactions can provide further solid basis to develop resistance in plants against *Fusarium* pathogens. Scientific advancement in technology such as development of new crop cultivars and transgenic plants resilient to *Fusarium* infections could also reduce crop and economic losses worldwide. Further, a deeper knowledge of *Fusarium* pathogens and its effect on plants can assist in better understanding ecosystems for sustainable agriculture and complete control of *Fusarium* pathogens.

Author details

Nadeem Iqbal[1,2]*, Riyazuddin Riyazuddin[1,3], Muhammad Nauman[4], Zalán Czékus[1], Malik Tahir Hayat[4], Péter Poór[1] and Attila Ördög[1]

1 Faculty of Science and Informatics, Department of Plant Biology, University of Szeged, Szeged, Hungary

2 School of Environmental Sciences, University of Szeged, Szeged, Hungary

3 Institute of Plant Biology, Biological Research Centre, Szeged, Hungary

4 Department of Environmental Sciences, COMSATS University Islamabad, Abbottabad, Pakistan

*Address all correspondence to: nadeemiqbal814@gmail.com

References

[1] FAO, IFAD, UNICEF, WFP, WHO. The State of Food Security and Nutrition in the World. Building Climate Resilience for Food Security and Nutrition. Rome. Italy: FAO; 2018

[2] Godfray HCJ, Beddington JR, Crute IR, Haddad L, Lawrence D, Muir JF, et al. Food security: The challenge of feeding 9 billion people. Science. 2010;**327**:812-818

[3] Cao Y, Zhang Z, Ling N, Yuan Y, Zheng X, Shen B, et al. *Bacillus subtilis* SQR 9 can control *Fusarium* wilt in cucumber by colonizing plant roots. Biology and Fertility of Soils. 2011;**47**:495-506

[4] Mazzola M, Freilich S. Prospects for biological soil borne disease control: Application of indigenous versus synthetic microbiomes. Phytopathology. 2017;**107**:256-263

[5] Duan C, Qin Z, Yang Z, Li W, Sun S, Zhu Z, et al. Identification of pathogenic *Fusarium* spp. causing maize ear rot and potential mycotoxin production in China. Toxins. 2016;**8**:186

[6] McGovern RJ. Management of tomato diseases caused by *Fusarium oxysporum*. Crop Protection. 2015;**73**:78-92

[7] Gu Q, Yang Y, Yuan Q, Shi G, Wu L, Lou Z, et al. Bacillomycin D produced by *bacillus amyloliquefaciens* is involved in the antagonistic interaction with the plant-pathogenic fungus *Fusarium graminearum*. Applied and Environmental Microbiology. 2017;**83**:e01075-e01017

[8] Garcia D, Barros G, Chulze S, Ramos AJ, Sanchis V, Marín S. Impact of cycling temperatures on *Fusarium verticillioides* and *Fusarium graminearum* growth and mycotoxins production in soybean. Journal of the Science of Food and Agriculture. 2012;**92**:2952-2959

[9] Tekle S, Skinnes H, Bjørnstad A. The germination problem of oat seed lots affected by *Fusarium* head blight. European Journal of Plant Pathology. 2013;**135**:147-158

[10] Hofgaard IS, Aamot HU, Torp T, Jestoi M, Lattanzio VMT, Klemsdal SS, et al. Associations between *Fusarium* species and mycotoxins in oats and spring wheat from farmers' fields in Norway over a six-year period. World Mycotoxin Journal. 2016;**9**:365-378

[11] Hjelkrem AGR, Torp T, Brodal G, Aamot HU, Strand E, Nordskog B, et al. DON content in oat grains in Norway related to weather conditions at different growth stages. European Journal of Plant Pathology. 2017;**148**:577-594

[12] Packa D, Załuski D, Graban Ł, Lajszner W, Hościk M. Reakcja diploidalnych, tetraploidalnych I heksaploidalnych pszenic na inokulację *Fusarium culmorum* (W.G.Smith) Sacc. Polish Journal of Agronomy. 2013;**12**:38-48

[13] Salgado JD, Madden LV, Paul PA. Efficacy and economics of integrating in-field and harvesting strategies to manage *Fusarium* head blight of wheat. Plant Disease. 2014;**98**:1407-1421

[14] Mielniczuk E, Skwaryło-Bednarz B. *Fusarium* head blight, mycotoxins and strategies for their reduction. Agronomy. 2020;**10**(4):509

[15] Iqbal N, Czékus Z, Poór P, Ördög A. Plant defence mechanisms against

mycotoxin Fumonisin B1. Chemico-Biological Interactions. 2021;**343**:109494

[16] Iqbal N, Czékus Z, Ördög A, Poór P. The main fungal pathogens and defence-related hormonal signalling in crops. In: Khan MIR, Singh A, Poór P, editors. Plant Hormones in Crop Improvement. Cambridge Academic Press; 2023. pp. 307-331

[17] Sadeghpour N, Asadi-Gharneh HA, Nasr-Esfahani M, Khankahdani HH, Golabadi M. Antioxidant enzymes associated with resistance to *Fusarium oxysporum* f. sp. melonis race 1.2 in melon. Physiological and Molecular Plant Pathology. 2022;**121**:101880

[18] Seiber JN, Coats J, Duke SO, Gross AD. Biopesticides: State of the art and future opportunities. Journal of Agricultural and Food Chemistry. 2014;**62**:11613-11619

[19] Veliz EA, Martinez-Hidalgo P, Hirsch AM. Chitinase producing bacteria and their role in biocontrol. AIMS Microbiology. 2017;**3**:689-705

[20] Babadoost M. Fusarium: Historical and continued importance. In: Fusarium AT, editor. Plant Diseases, Pathogen Diversity, Genetic Diversity, Resistance and Molecular Markers. London, UK: Intech Open; 2018

[21] Mohamed Zubi WS, Mohd MH, Mohamed Nor NMI, Zakaria L. *Fusarium* species in mangrove soil in northern peninsular Malaysia and the soil physico-chemical properties. Microorganisms. 2021;**9**:497

[22] Summerell BA, Laurence MH, Liew ECY, Leslie JF. Biogeography and phylogeography of *Fusarium*: A review. Fungal Diversity. 2010;**44**:3-13

[23] Summerell BA, Leslie JF, Liew ECY, Laurence MH, Bullock S, Petrovic T, et al. *Fusarium* species associated with plants in Australia. Fungal Diversity. 2011;**46**:1-27

[24] Iqbal N, Czékus Z, Ördög A, Poór P. Fusaric acid-evoked oxidative stress affects plant defence system by inducing biochemical changes at subcellular level. Plant Cell Reports. 2024;**43**(1):2

[25] Desai S, Dubey SC, Prasad RD. Impacts of climate change on *Fusarium* species Vis-à-Vis adaptation strategies. Indian Phytopathology. 2020;**73**(4):593-603

[26] Iqbal N, Czékus Z, Angeli C, Bartók T, Poór P, Ördög A. Fumonisin B1-induced oxidative burst perturbed photosynthetic activity and affected antioxidant enzymatic response in tomato plants in ethylene-dependent manner. Journal of Plant Growth Regulation. 2023;**42**(3):1865-1878

[27] Rampersad SN. Pathogenomics and management of *Fusarium* diseases in plants. Pathogens. 2020;**9**(5):340

[28] Incremona ME, Gonzalez M, Pioli RN, Salinas A. Infection of maize silks by a native *Fusarium* (*Fusarium graminearum*) isolate in Argentina. Chilean Journal of Agricultural and Animal Science. 2014;**30**:203-211

[29] Perincherry L, Lalak-Kańczugowska J, Stępień Ł. *Fusarium*-produced mycotoxins in plant-pathogen interactions. Toxins. 2019;**11**(11):664

[30] Ekwomadu TI, Akinola SA, Mwanza M. *Fusarium* mycotoxins, their metabolites (free, emerging, and masked), food safety concerns, and health impacts. International Journal of Environmental Research and Public Health. 2021;**18**(22):11741

[31] Eskola M, Kos G, Elliott CT, Hajšlová J, Mayar S, Krska R. Worldwide

contamination of food-crops with mycotoxins: Validity of the widely cited 'FAO estimate' of 25%. Critical Reviews in Food Science and Nutrition. 2020;**60**:2773-2789

[32] Reddy KRN, Nurdijati SB, Salleh B. An overview of plant-derived products on control of mycotoxigenic fungi and mycotoxins. Asian Journal of Plant Sciences. 2010;**9**:126

[33] Nag P, Paul S, Shriti S, Das S. Defence response in plants and animals against a common fungal pathogen, *Fusarium oxysporum*. Current Research in Microbial Sciences. 2022;**3**:100135

[34] Li CY, Deng GM, Yang J, Viljoen A, Jin Y, Kuang RB, et al. Transcriptome profiling of resistant and susceptible Cavendish banana roots following inoculation with *Fusarium oxysporum* f. sp. cubense tropical race 4. BMC Genomics. 2012;**13**(1):1-11

[35] Muthamilarasan M, Prasad M. Plant innate immunity: An updated insight into defence mechanism. Journal of Biosciences. 2013;**38**:433-449

[36] Wojtasik W, Preisner M, Boba A, Kostyn K, Dymińska L, Hanuza J, et al. Rearrangement of cell wall polymers in flax infected with a pathogenic strain of *Fusarium culmorum*. Physiological and Molecular Plant Pathology. 2020;**110**:101461

[37] Anisimova OK, Shchennikova AV, Kochieva EZ, Filyushin MA. Pathogenesis-related genes of PR1, PR2, PR4, and PR5 families are involved in the response to *Fusarium* infection in garlic (*Allium sativum* L.). International Journal of Molecular Sciences. 2021;**22**(13):6688

[38] Beccari G, Hao G, Liu H. *Fusarium* pathogenesis: Infection mechanisms and disease progression in host plants. Frontiers in Plant Science. 2022;**13**:1020404

[39] Husaini AM, Sakina A, Cambay SR. Host–pathogen interaction in *Fusarium oxysporum* infections: Where do we stand? Molecular Plant-Microbe Interactions. 2018;**31**(9):889-898

[40] Ridenour JB, Bluhm BH. The HAP complex in *Fusarium verticillioides* is a key regulator of growth, morphogenesis, secondary metabolism, and pathogenesis. Fungal Genetics and Biology. 2014;**69**:52-64

[41] Li M, Xie L, Wang M, Lin Y, Zhong J, Zhang Y, et al. FoQDE2-dependent milRNA promotes *Fusarium oxysporum* f. sp. cubense virulence by silencing a glycosyl hydrolase coding gene expression. PLoS Pathogens. 2022;**18**(5):e1010157

[42] Perez-Nadales E, Di Pietro A. The transmembrane protein Sho1 cooperates with the mucin Msb2 to regulate invasive growth and plant infection in *Fusarium oxysporum*. Molecular Plant Pathology. 2015;**16**(6):593-603

[43] Ploetz RC. *Fusarium* wilt of banana. Phytopathology. 2015;**105**:1512-1521

[44] Tiwari N, Ahmed S, Kumar S, Sarker A. *Fusarium* wilt: A killer disease of lentil. In: Asku T, editor. *Fusarium*- Plant Diseases, Pathogen Diversity, Genetic Diversity, Resistance and Molecular Markers. London, UK, Rijeka, Croatia: IntechOpen; 2018

[45] Popescu S, Boldura OM, Borozan A, Madosa E. Molecular detection of *Fusarium* infections in wheat: A measure of quality assessment. Journal on Processing and Energy in Agriculture. 2022;**26**(1):27-29

[46] Awuchi CG, Ondari EN, Eseoghene IJ, Twinomuhwezi H, Amagwula IO, Morya S. Fungal growth and mycotoxins production: Types, toxicities, control strategies, and detoxification. In: Sultan S, GKS S, editors. Fungal Reproduction and Growth. London: IntechOpen; 2021. p. 100207

[47] Stępień Ł, Koczyk G, Waśkiewicz A. FUM cluster divergence in fumonisins-producing *Fusarium* species. Fungal Biology. 2011;**115**(2):112-123

[48] Gelineau-van WJ. Fumonisins. In: Gupta RC, editor. Reproductive and Developmental Toxicology. Cambridge: Academic Press; 2022. pp. 955-981

[49] Tiwari R, Shukla SK, Jaiswal VP, Tiwari RK. Pokkah boeng disease of sugarcane: Current status and opportunities. 2020;**12**(1):1-6

[50] Fernandes LB, Ghag SB. Molecular insights into the jasmonate signalling and associated defence responses against wilt caused by *Fusarium oxysporum*. Plant Physiology and Biochemistry. 2022;**174**:22-34

[51] Niehaus EM, Díaz-Sánchez V, von Bargen KW, Kleigrewe K, Humpf HU, Limón MC, et al. Fusarins and fusaric acid in fusaria. In: Martín JF, García-Estrada C, Zeilinger S, editors. Biosynthesis and Molecular Genetics of Fungal Secondary Metabolites. New York: Springer; 2014. pp. 239-262

[52] Stępień Ł, Lalak-Kańczugowska J, Witaszak N, Urbaniak M. *Fusarium* secondary metabolism biosynthetic pathways: So close but so far away. In: Mérillon JM, Ramawat KG, editors. Co-Evolution of Secondary Metabolites. Cham: Springer Nature; 2020. pp. 211-247

[53] Selim ME, El-Gammal NA. Role of fusaric acid mycotoxin in pathogensis process of tomato wilt disease caused by *Fusarium oxysporum*. Journal of Bioprocessing and Biotechniques. 2015;**5**:1

[54] Singh VK, Upadhyay RS. Fusaric acid induced cell death and changes in oxidative metabolism of *Solanum lycopersicum* L. Botanical Studies. 2014;**55**:66

[55] Iqbal N, Czékus Z, Ördög A, Poór P. Ethylene-dependent effects of fusaric acid on the photosynthetic activity of tomato plants. Photosynthetica. 2021;**59**(2):337-348

[56] Singh VK, Singh HB, Upadhyay RS. Role of fusaric acid in the development of '*Fusarium* wilt' symptoms in tomato: Physiological, biochemical and proteomic perspectives. Plant Physiology and Biochemistry. 2017;**118**:320-332

[57] Bertuzzi T, Giorni P, Rastelli S, Vaccino P, Lanzanova C, Locatelli S. Co-occurrence of moniliformin and regulated *Fusarium* toxins in maize and wheat grown in Italy. Molecules. 2020;**25**(10):2440

[58] Urbaniak M, Stepie'n Ł, Uhlig S. Evidence for naturally produced Beauvericins containing *N*-methyl-tyrosine in *Hypocreales* Fungi. Toxins. 2019;**11**:182

[59] Mallebrera B, Prosperini A, Font G, Ruiz MJ. In vitro mechanisms of Beauvericin toxicity: A review. Food and Chemical Toxicology. 2018;**111**:537-545

[60] Poppenberger B, Berthiller F, Lucyshyn D, Sieberer T, Schuhmacher R, Krska R, et al. Detoxification of the *Fusarium* mycotoxin deoxynivalenol by a UDP-glucosyltransferase from *Arabidopsis thaliana*. Journal of Biological Chemistry. 2003;**278**:47905-47914

[61] Boddu J, Cho S, Muehlbauer GJ. Transcriptome analysis of trichothecene-induced gene expression in barley. Molecular Plant-Microbe Interactions. 2007;**20**:1364-1375

[62] Huang J, Pang C, Fan S, Song M, Yu J, Wei H, et al. Genome-wide analysis of the family 1 glycosyltransferases in cotton. Molecular Genetics and Genomics. 2015;**290**:1805-1818

[63] ter Beek J, Guskov A, Slotboom DJ. Structural diversity of ABC transporters. Journal of General Physiology. 2014;**143**(4):419-435

[64] Van Schie CC, Takken FL. Susceptibility genes 101: How to be a good host. Annual Review of Phytopathology. 2014;**52**:551-581

[65] Pereyra MG, Sulyok M, Baralla V, Dalcero AM, Krska R, Chulze S, et al. Evaluation of zearalenone, α-zearalenol, β-zearalenol, zearalenone 4-sulfate and β-zearalenol 4-glucoside levels during the ensiling process. World Mycotoxin Journal. 2014;**7**(3):291-295

[66] Ohsato S, Ochiai-Fukuda T, Nishiuchi T, Takahashi-Ando N, Koizumi S, Hamamoto H, et al. Transgenic rice plants expressing trichothecene 3-*O*-acetyltransferase show resistance to the *Fusarium* phytotoxin deoxynivalenol. Plant Cell Reports. 2007;**26**:531-538

[67] Li X, Shin S, Heinen S, Dill-Macky R, Berthiller F, Nersesian N, et al. Transgenic wheat expressing a barley UDP-glucosyltransferase detoxifies deoxynivalenol and provides high levels of resistance to *Fusarium graminearum*. Molecular Plant-Microbe Interactions. 2015;**28**:1237-1246

[68] Perochon A, Váry Z, Malla KB, Halford NG, Paul MJ, Doohan FM. The wheat SnRK1α family and its contribution to *Fusarium* toxin tolerance. Plant Science. 2019;**288**:110217

[69] Arie T. *Fusarium* diseases of cultivated plants, control, diagnosis, and molecular and genetic studies. Journal of Pesticide Science. 2019;**44**(4):275-281

[70] Zakaria L. *Fusarium* species associated with diseases of major tropical fruit crops. Horticulturae. 2023;**9**(3):322

[71] Khan N, Maymon M, Hirsch AM. Combating *Fusarium* infection using *Bacillus*-based antimicrobials. Microorganisms. 2017;**5**(4):75

[72] Fourie G, Steenkamp ET, Ploetz RC, Gordon TR, Viljoen A. Current status of the taxonomic position of *Fusarium oxysporum* formae specialis cubense within the *Fusarium oxysporum* complex. Infection, Genetics and Evolution. 2011;**11**:533-542

[73] De Sain M, Rep M. The role of pathogen-secreted proteins in fungal vascular wilt diseases. International Journal of Molecular Sciences. 2015;**16**:23970-23993

[74] Mendel Z, Protosav A, Sharon M, Zveibil A, Ben Yahuda S, O'Donnell K, et al. An Asian ambrosia beetle *Euwallacea fornicatus* and its novel symbiotic fungus *Fusarium* sp. Pose a serious threat to Israeli avocado industry. Phytoparasitica. 2012;**40**:235-238

[75] Yadeta KA, Thomma BP. The xylem as battleground for plant hosts and vascular wilt pathogens. Frontiers in Plant Science. 2013;**4**:97

[76] Kazan K, Gardiner DM. Transcriptomics of cereal–*Fusarium* graminearum interactions: What we have learned so far. Molecular Plant Pathology. 2018;**19**(3):764-778

[77] Del Ponte EM, Valent B, Bergstrom GC. A special issue on *Fusarium* head blight and wheat blast. Tropical Plant Pathology. 2017;**42**:143-145

[78] Dweba CC, Figlan S, Shimelis HA, Motaung TE, Sydenham S, Mwadzingeni L, et al. *Fusarium* head blight of wheat: Pathogenesis and control strategies. Crop Protection. 2017;**91**:114-122

[79] Seo Y, Kim YH. Potential reasons for prevalence of *Fusarium* wilt in oriental melon in Korea. Plant Pathology Journal. 2017;**33**:249-263

[80] Ma YT, Fan HF, Gao YQ, Li H, Zhang AL, Gao JM. Natural products as sources of new fungicides (I): Synthesis and antifungal activity of acetophenone derivatives against phytopathogenic fungi. Chemical Biology and Drug Research. 2013;**81**:545-552

[81] Everts KL, Himmelstein JC. *Fusarium* wilt of watermelon: Towards sustainable management of are-emerging plant disease. Crop Protection. 2015;**73**:93-99

[82] Heydari A, Pessarakli M. A review on biological control of fungal plant pathogens using microbial antagonists. Journal of Biological Sciences. 2010;**10**:273-290

[83] Bakkera MG, Gloverb JD, Maib JG, Kinkela LL. Plant community effects on the diversity and pathogen suppressive activity of soil *streptomycetes*. Applied Soil Ecology. 2010;**46**:35-42

[84] Dunlap CA, Schisler DA, Bowman MJ, Rooney AP. Genomic analysis of *Bacillus subtilis* OH 131.1 and co-culturing with *Cryptococcus flavescens* for control of *Fusarium* head blight. Plant Gene. 2015;**2**:1-9

[85] Zalila-Kolsi I, Mahmoud AB, Ali H, Sellami S, Nasfi Z, Tounsi S, et al. Antagonist effects of *bacillus* spp. strains against *Fusarium graminearum* for protection of durum wheat (*Triticum turgidum* L. sub sp. *durum*). Microbiological Research. 2016;**192**:148-158

[86] Atanasova-Penichon V, Barreau C, Richard-Forget F. Antioxidant secondary metabolites in cereals: Potential involvement in resistance to *Fusarium* and mycotoxin accumulation. Frontiers in Microbiology. 2016;**7**:566

[87] Chrpová J, Orsák M, Martinek P, Lachman J, Trávníčková M. Potential role and involvement of antioxidants and other secondary metabolites of wheat in the infection process and resistance to *Fusarium* spp. Agronom. 2021;**11**(11):2235

[88] Anthony KK, George DS, Baldev Singh HK, Fung SM, Santhirasegaram V, Razali Z, et al. Reactive oxygen species activity and antioxidant properties of *Fusarium* infected bananas. Journal of Phytopathology. 2017;**165**(4):213-222

[89] Zehra A, Meena M, Dubey MK, Aamir M, Upadhyay RS. Synergistic effects of plant defence elicitors and *Trichoderma harzianum* on enhanced induction of antioxidant defence system in tomato against *Fusarium* wilt disease. Botanical Studies. 2017;**58**(1):1-14

[90] Hashem A, Akhter A, Alqarawi AA, Singh G, Almutairi KF, Abd_Allah EF. Mycorrhizal fungi induced activation of tomato defence system mitigates *Fusarium* wilt stress. Saudi Journal of Biological Sciences. 2021;**28**(10):5442-5450

[91] Gao T, Zhang Y, Shi J, Mohamed SR, Xu J, Liu X. The antioxidant guaiacol exerts fungicidal activity against fungal

growth and deoxynivalenol production in *Fusarium graminearum*. Frontiers in Microbiology. 2021;**12**:762844

[92] Sun S, Yang Z, Song Z, Wang N, Guo N, Niu J, et al. Silicon enhances plant resistance to *Fusarium* wilt by promoting antioxidant potential and photosynthetic capacity in cucumber (*Cucumis sativus* L.). Frontiers in Plant Science. 2022;**13**:1011859

[93] Matić M, Vuković R, Vrandečić K, Štolfa Čamagajevac I, Ćosić J, Vuković A, et al. Oxidative status and antioxidative response to *Fusarium* attack and different nitrogen levels in winter wheat varieties. Plants. 2021;**10**(4):611

[94] Abdelaziz AM, Attia MS, Salem MS, Refaay DA, Alhoqail WA, Senousy HH. *Cyanobacteria*-mediated immune responses in pepper plants against *Fusarium* wilt. Plants. 2022;**11**(15):2049

[95] Batnini M, Lopez-Gomez M, Palma F, Haddoudi I, Kallala N, Zribi K, et al. *Sinorhizobium* spp inoculation alleviates the effect of *Fusarium oxysporum* on *Medicago truncatula* plants by increasing antioxidant capacity and sucrose accumulation. Applied Soil Ecology. 2020;**150**:103458

[96] Balmer D, Flors V, Glauser G, Mauch-Mani B. Metabolomics of cereals under biotic stress: Current knowledge and techniques. Frontiers in Plant Science. 2013;**4**:82

[97] Picot A, Atanasova-Pénichon V, Pons S, Marchegay G, Barreau C, Pinson-Gadais L, et al. Maize kernel antioxidants and their potential involvement in *Fusarium* ear rot resistance. Journal of Agricultural and Food Chemistry. 2013;**61**:3389-3395

[98] Etzerodt T, Gislum R, Laursen BB, Heinrichson K, Gregersen PL, Jørgensen LN, et al. Correlation of deoxynivalenol accumulation in Fusarium-infected winter and spring wheat cultivars with secondary metabolites at different growth stages. Journal of Agricultural and Food Chemistry. 2016;**64**:4545-4555

[99] Kumaraswamy GK, Kushalappa AC, Choo TM, Dion Y, Rioux S. Differential metabolic response of barley genotypes, varying in resistance, to trichothecene-producing and-nonproducing (tri5-) isolates of *Fusarium graminearum*. Plant Pathology. 2012;**61**:509-521

[100] Ferruz E, Atanasova-Pénichon V, Bonnin-Verdal MN, Marchegay G, Pinson-Gadais L, Ducos C, et al. Effects of phenolic acids on the growth and production of T-2 and HT-2 toxins by *Fusarium langsethiae* and *F. Sporotrichioides*. Molecules. 2016;**21**:449

[101] Katagiri F, Tsuda K. Understanding the plant immune system. Molecular Plant Microbe Interactions. 2010;**23**(12):15311536

[102] Couto D, Zipfel C. Regulation of pattern recognition receptor signalling in plants. Nature Reviews Immunology. 2016;**16**:537-552

[103] Verma V, Ravindran P, Kumar PP. Plant hormone-mediated regulation of stress responses. BMC Plant Biology. 2016;**16**(1):1-10

[104] Ghorbel M, Brini F. Hormone mediated cell signalling in plants under changing environmental stress. Plant Gene. 2021;**28**:100335

[105] Yu SG, Kim JH, Cho NH, Oh TR, Kim WT. *Arabidopsis* RING E3 ubiquitin ligase JUL1 participates in ABA-mediated microtubule depolymerization, stomatal closure, and tolerance response to

drought stress. The Plant Journal. 2020;**103**(2):824-842

[106] Torres-Vera R, García JM, Pozo MJ, López-Ráez JA. Do strigolactones contribute to plant defence? Molecular Plant Pathology. 2014;**15**(2):211-216

[107] Takatsuji H, Jiang CJ. Plant hormone crosstalks under biotic stresses. In: Tran LS, Pal S, editors. Phytohormones: A Window to Metabolism, Signalling and Biotechnological Applications. New York: Springer; 2014. pp. 323-350

[108] Qi PF, Johnston A, Balcerzak M, Rocheleau H, Harris LJ, Long XY, et al. Effect of salicylic acid on *Fusarium graminearum*, the major causal agent of *Fusarium* head blight in wheat. Fungal Biology. 2012;**116**(3):413-426

[109] Makandar R, Nalam V, Chaturvedi R, Jeannotte R, Sparks AA, Shah J. Involvement of interaction between salicylic acid and jasmonate signalling pathways in *Arabidopsis* interaction with *Fusarium graminearum*. Molecular Plant Microbe Interactions. 2010;**23**:861-870

[110] Yang YX, J Ahammed G, Wu C, Fan SY, Zhou YH. Crosstalk among jasmonate, salicylate and ethylene signalling pathways in plant disease and immune responses. Current Protein and Peptide Science. 2015;**16**(5):450-461

[111] Xue RF, Wu J, Wang LF, Blair MW, Wang XM, De Ge W, et al. Salicylic acid enhances resistance to *Fusarium oxysporum* f. sp. phaseoli in common beans (*Phaseolus vulgaris L.*). Journal of Plant Growth Regulation. 2013;**33**:470-476

[112] Djamei A, Schipper K, Rabe F, Ghosh A, Vincon V, Kahnt J, et al. Metabolic priming by a secreted fungal effector. Nature. 2011;**478**:395-398

[113] Han X, Kahmann R. Manipulation of phytohormone pathways by effectors of filamentous plant pathogens. Frontiers in Plant Science. 2019;**10**:822

[114] Chakraborty N. Salicylic acid and nitric oxide cross-talks to improve innate immunity and plant vigor in tomato against *Fusarium oxysporum* stress. Plant Cell Reports. 2021;**40**(8):1415-1427

[115] Guan Y, He X, Wen D, Chen S, Chen F, Chen F, et al. *Fusarium oxysporum* infection on root elicit aboveground terpene production and salicylic acid accumulation in *Chrysanthemum morifolium*. Plant Physiology and Biochemistry. 2022;**190**:11-23

[116] Taheri P, Tarighi S. Riboflavin induces resistance in rice against *Rhizoctonia solani* via jasmonate-mediated priming of phenylpropanoid pathway. Journal of Plant Physiology. 2010;**167**(3):201-208

[117] Di X, Gomila JO, Takken FL. Involvement of salicylic acid, ethylene and jasmonic acid signalling pathways in the susceptibility of tomato to *Fusarium oxysporum*. Molecular Plant Pathology. 2017;**18**(7):1024-1035

[118] Qi PF, Balcerzak M, Rocheleau H, Leung W, Wei YM, Zheng YL, et al. Jasmonic acid and abscisic acid play important roles in host–pathogen interaction between *Fusarium graminearum* and wheat during the early stages of *Fusarium* head blight. Physiological and Molecular Plant Pathology. 2016;**93**:39-48

[119] Siciliano I, Amaral Carneiro G, Spadaro D, Garibaldi A, Gullino ML. Jasmonic acid, abscisic acid, and salicylic

acid are involved in the phytoalexin responses of rice to *Fusarium fujikuroi*, a high gibberellin producer pathogen. Journal of Agricultural and Food Chemistry. 2015;**63**(37):8134-8142

[120] Anver S, Tsuda K. Ethylene and plant immunity. In: Wen CK, editor. Ethylene in Plants. Dordrecht: Springer; 2015. p. 205221

[121] Pantelides IS, Tjamos SE, Pappa S, Kargakis M, Paplomatas EJ. The ethylene receptor ETR1 is required for *Fusarium oxysporum* pathogenicity. Plant Pathology. 2013;**62**:1302-1309

[122] Chen X, Steed A, Travella S, Keller B, Nicholson P. *Fusarium graminearum* exploits ethylene signalling to colonize dicotyledonous and monocotyledonous plants. New Phytologist. 2009;**182**:975-983

[123] Dinolfo MI, Castañares E, Stenglein SA. *Fusarium*-plant interaction: State of the art-a review. Czech Academy of Agricultural Sciences; Plant Protection Science. 2017;**53**(2):61-70

[124] Cakir B, Gül A, Yolageldi L, Özaktan H. Response to *Fusarium oxysporum* f.sp. radices-lycopersici in tomato roots involves regulation of SA- and ET-responsive gene expressions. European Journal of Plant Pathology. 2014;**139**:379-391

[125] Luu VT, Schuck S, Kim SG, Weinhold A, Baldwin IT. Jasmonic acid signalling mediates resistance of the wild tobacco *Nicotiana attenuata* to its native *Fusarium*, but no *Alternaria*, fungal pathogens. Plant, Cell and Environment. 2015;**38**:572-584

[126] Asano T, Kimura M, Nishiuchi T. The defence response in *Arabidopsis thaliana* against *Fusarium sporotrichioides*. Proteome Science. 2012;**10**:61

[127] Gao CS, Kou XJ, Li HP, Zhang JB, Saad ASI, Liao YC. Inverse effects of *Arabidopsis* NPR1 gene on *Fusarium* seedling blight and *Fusarium* head blight in transgenic wheat. Plant Pathology. 2012;**62**:383-392

[128] Haidoulis JF, Nicholson P. Different effects of phytohormones on *Fusarium* head blight and *Fusarium* root rot resistance in *Brachypodium distachyon*. Journal of Plant Interactions. 2020;**15**(1):335-344

[129] Law JWF, Ser HL, Khan TM, Chuah LH, Pusparajah P, Chan KG, et al. The potential of *Streptomyces* as biocontrol agents against the rice blast fungus, *Magnaporthe oryzae* (*Pyricularia oryzae*). Frontiers in Microbiology. 2017;**8**:3

[130] Donato V, Ayala FR, Cogliati S, Bauman C, Costa JB, Lenini C, et al. *Bacillus subtilis* biofilm extends *Caenorhabditis elegans* longevity through downregulation of the insulin-like signalling pathway. Nature Communications. 2017;**8**:14332

[131] Allard-Massicotte R, Tessier L, Lecuyer F, Lakshmanan V, Lucier JF, Garneau D, et al. *Bacillus subtilis* early colonization of *Arabidopsis thaliana* roots involve multiple chemotaxis receptors. mBio. 2016;7:e01664-e01616

[132] Heidarzadeh N, Baghaee-Ravari S. Application of *Bacillus pumilus* as a potential biocontrol agent of *Fusarium* wilt of tomato. Archives of Phytopathology and Plant Protection. 2015;**48**:13-16

[133] Swiontek BM, Jankiewicz U, Burkowska A, Walczak M. Chitinolytic microorganisms and their possible application in environmental protection. Current Microbiology. 2014;**68**:71-81

[134] Zhao Y, Selvaraj JN, Xing F, Zhou L, Wang Y, Song H, et al. Antagonistic action of *Bacillus subtilis* strain SG6 on *Fusarium graminearum*. PLoS One. 2014;**9**:e92486

[135] Gomaa EZ. Chitinase production by *Bacillus thuringiensis* and *Bacillus licheniformis:* Their potential in antifungal biocontrol. Journal of Microbiology. 2012;**50**:103-111

[136] Akram W, Anjum T, Ali B. Phenylacetic acid is ISR determinant produced by *Bacillus fortis* IAGS162, which involves extensive re-modulation in metabolomics of tomato to protect against *Fusarium* wilt. Frontiers in Plant Science. 2016;**7**:498

[137] Patil S, Sriram S. Biological control of *Fusarium* wilt in crop plants using non-pathogenic isolates of *Fusarium* species. Indian Phytopathology. 2020;**73**(1):11-19

[138] Thangavelu R, Mustaffa MM. Current advances in the Fusarium wilt disease management in banana with emphasis on biological control. In: Cumagun CJ, editor. Plant Pathology. London: IntechOpen; 2012. pp. 273-288

[139] Wojtasik W, Kulma A, Namys K, Preisner M, Szopa J. Polyamine metabolism in flax in response to treatment with pathogenic and non-pathogenic *Fusarium* strains. Frontiers in Plant Science. 2015;**6**:291

[140] Abderrahmane S, Merzoug A, Taleb M. The use of non-pathogenic *Fusarium oxysporum* as a biocontrol against *Fusarium* wilt of date palm (Bayoud disease). In: Proceedings of International Workshop on "Sustainability of Phoenici" Culture Production Systems in Algeria. Biskra. Algeria: University of Biskra; 2016. p. 67

[141] Nawar LS. Phytochemical and SDS-dissociated proteins of pathogenic and non-pathogenic *Fusarium oxysporum* isolates. International Journal of ChemTech Researh. 2016;**9**(6):165-172

[142] Podgórska-Kryszczuk I, Solarska E, Kordowska-Wiater M. Biological control of *Fusarium culmorum*, *Fusarium graminearum* and *Fusarium poae* by antagonistic yeasts. Pathogens. 2022;**11**(1):86

[143] Seepe HA, Lodama KE, Sutherland R, Nxumalo W, Amoo SO. In vivo antifungal activity of south African medicinal plant extracts against *Fusarium* pathogens and their phytotoxicity evaluation. Plants. 2020;**9**:1668

[144] Sultana S, Akhtar N, Asif HM. Phytochemical screening and antipyretic effects of hydro-methanol extract of *Melia azedarach* leaves in rabbits. Bangladesh Journal of Pharmacology. 2013;**8**:214-217

[145] Seepe HA, Nxumalo W, Amoo SO. Natural products from medicinal plants against phytopathogenic *Fusarium* species: Current research endeavours, challenges and prospects. Molecules. 2021;**26**(21):6539

[146] Mogashoa MM, Masoko P, Eloff JN. Different *Combretum molle* (Combretaceae) leaf extracts contain several different antifungal and antibacterial compounds. South African Journal of Botony. 2019;**126**:322-327

[147] Seepe HA, Amoo SO, Nxumalo W, Adeleke RA. Antifungal activity of medicinal plant extracts for potential management of *Fusarium* pathogens. Research on Crops. 2019;**20**:399-406

[148] Seepe HA, Amoo SO, Nxumalo W, Adeleke RA. Sustainable use of thirteen south African medicinal plants for the

management of crop diseases caused by *Fusarium* species–An in vitro study. South African Journal of Botony. 2020;**130**:456-464

[149] Li KM, Dong X, Ma YN, Wu ZN, Yan YM, Cheng YX. Antifungal coumarins and lignans from *Artemisia annua*. Fitoterapia. 2019;**134**:323-328

[150] He R, Guo D, Huang Z, Kong Y, Ji C, Gu J, et al. Systematic investigation of stereochemistry, stereoselective bioactivity, and antifungal mechanism of chiral triazole fungicide metconazole. Science of the Total Environment. 2021;**784**:147194

[151] Solanki MK, Abdelfattah A, Britzi M, Zakin V, Wisniewski M, Droby S, et al. Shifts in the composition of the microbiota of stored wheat grains in response to fumigation. Frontiers in Microbiology. 2019;**10**:1098

[152] Juodeikiene G, Bartkiene E, Cernauskas D, Cizeikiene D, Zadeike D, Lele V, et al. Antifungal activity of lactic acid bacteria and their application for *Fusarium* mycotoxins reduction in malting wheat grains. LWT-Food Science Technology. 2018;**89**:307-314

[153] Omotayo OP, Omotayo AO, Mwanza M, Babalola OO. Prevalence of mycotoxins and their consequences on human health. Toxicological Research. 2019;**35**:1-7

Chapter 5

Fusarium Species Responsible for Tomato Diseases and Mycotoxin Contamination and Biocontrol Opportunities

Daniela Simões and Eugénia de Andrade

Abstract

For many years, *Fusarium* species have been known as one of the most common pathogens, causing disease and producing mycotoxins in many host species both on-field and postharvest. Tomato is among the most relevant hosts due to its economic and nutritional relevance, its plasticity to be cultivated under diverse soils and climates, and consumed fresh or processed. The most common pathogenic *Fusarium* species are *Fusarium oxysporum f. spp. lycopersici, Fusarium oxysporum* f. sp. *radicis-lycopersici,* and *Fusarium solani*. However, the species presence and prevalence depend on the globe region, and other species can be found such as *F. semitectum, F. oxysporum, F. equiseti, F. falciforme,* or *F. striatum*. Most of these species' strains are also mycotoxigenic and can potentially contaminate tomatoes and tomato-based products with several mycotoxins. Some cases of mycotoxin contamination on tomatoes were reported and caused by different fungal species. Emerging *Fusarium* mycotoxins have recently been reported and gained high interest due to their increasing frequency. These mycotoxins, still not deeply studied, may constitute high-risk factors for human and animal health. This chapter is dedicated to the most relevant *Fusarium* spp. affecting tomato crops and the consequences of consuming mycotoxin-contaminated fresh tomatoes or industrially processed and describes some promising biocontrol measures.

Keywords: *Fusarium* spp., tomato, fungal diseases, mycotoxins, plant health, biocontrol

1. Introduction

Tomato (*Lycopersicon esculentum* L.) is one of the most popular vegetal crops in the world. Originally from the Andean region of Colombia, Peru, and Bolivia, it was domesticated in Mexico and introduced in Europe approximately 500 years ago [1]. Despite in South America, the fruit was considered edible, and Mexicans used it for different culinary applications; in Europe, due to its content in alkaloids, tomato plant was first used as ornamental and medicinal [1, 2]. Only 350 years ago, it was introduced in Spanish cuisine and Italian cuisine, followed by a widespread consumption [1]. Nowadays, tomato is an essential crop in nearly half of the existing countries (114 countries) and contributing significantly to the national food supply of 155 countries, accounting for 16% of the total production of vegetable species in 2020, followed by onions (9%), cucumbers (8%), cabbages (6%) and eggplants (5%) [3].

According to FAO, the total tomato production in the world was 184.8 and 189.1 million metric tons (mT), in 2020 and 2021, respectively [4]. In these years, China was the top producer accounting for 35–36% of the worldwide harvest [3], while in Europe, Italy had the highest production (6,644,790 mT), followed by Spain (4,754,380 mT), Ukraine (2,444,880 mT), and Portugal (1,741,320 mT). In terms of yield per hectare, Portugal was the one with the highest yield (97.9 mT/ha) compared to the other three European countries, while the highest farm yield was achieved in the Netherlands (475.7 mT/ha), representing eight times the yield of China (59.2 mT/ha) [5].

However, despite decades of conventional breeding and selection, the tomato crop is still susceptible to over 200 diseases and disorders during its growing season in outdoor production or greenhouses [6–8]. Some of them are common to all the production regions, while others are specific to some countries or locations. These causal agents may be diverse, including bacteria, viruses, viroids, nematodes, and fungi [9], the last most common cause of destructive diseases and mycotoxin contamination [10, 11]. Soilborne and foliar fungal diseases are the major limiting factors for tomato production [11]. Furthermore, tomato fruits are highly susceptible to fungal contamination during transportation, processing, and storage (postharvest diseases). The fungi may enter the fruits through damaged skin, or other tissues, taking the fruit at the market to wilt and squash easily [12].

Fungi have diverse but important roles in different ecosystems, with some species being crucial for soil fertility or support for plant species development [13], growth, fitness, and tolerance to stress, pests, pathogens, or herbivory (endophytic fungi) [14], or even a food-source for animals and humans [15]. However, some fungal species can also cause a panoply of diseases in plants, animals, and humans [16]. In fact, 70% of around 20% of the crop losses are caused by fungal diseases, resulting in high economic costs [17, 18]. Furthermore, most of the fungal species can produce different primary and secondary metabolites that can have beneficial or harmful effects on the surrounding organisms, which may have a positive agricultural or medicinal value, or cause severe health issues in animals and humans, like mycotoxins do [19, 20]. The most critical mycotoxins to animal and human health are aflatoxins, fumonisins (FUM), ochratoxin A (OTA), deoxynivalenol (DON), zearalenone (ZEN), and ergot alkaloids, mainly produced by *Aspergillus, Penicillium, Fusarium, Stachybotrys*, and *Claviceps* species [21, 22].

The main fungal genera found in tomato plants and fruits are saprotrophic, being naturally present in soil and comprising species with an important role in soil fertility, decomposing organic matter, and facilitating carbon and nitrogen flow such as *Alternaria, Fusarium, Penicillium,* and *Aspergillus*, that can be easily found mainly in postharvested tomatoes [10, 23]. However, some of these fungi can also be highly pathogenic and capable to producing mycotoxins, constituting a high risk of yield and quality losses [16]. In fact, the *Alternaria* genus is responsible for a decrease of 50–80% of the crop yield, causing "Early Blight" (mainly *Alternaria solani*) and includes mycotoxigenic species that produce tenuazonic acid (TeA) and alterotoxins [10]. Also, *Fusarium* spp. can cause several diseases on tomato plants and is responsible for "tomato fruit rot" post-harvesting. Furthermore, *Fusarium* spp. have the ability to synthesize most of the major mycotoxins, including moniliformin (MON), FUM, DON and ZEN [12, 16, 24, 25], which could potentially leading to contamination of the fruits. In tomato crops, one can find other fungal genera responsible for diseases, such as the primary saprophytic *Septoria lycopersici*, causing "Septoria Leaf Spot", and *Botrytis cinerea*, causing "Gray Mold" [26–29], the obligate parasite Leveillula taurica (responsible for "Powdery Mildew"), the phytopatogenic fungi *Colletotrichum* spp. (Anthracnose) and *Verticillium* spp. (*verticillium* Wilt), and the pathogenic oomycete Phytophthora infestans [8–11]. Typical phytopathogenic fungi also infect tomato plants causing "Anthracnose" (*Colletotrichum* spp.), "*Verticillium* Wilt" (*Verticillium* spp.), "Powdery Mildew" (*Leveillula taurica* - an obligate fungal parasite), or "Late Blight" (*Phytophthora infestans* – a pathogenic oomycete), for example [8, 11]. The tomato industry relies on fresh and frozen tomatoes as raw materials being "Gray Mold" and "Soft Rot" the most important postharvest disease of tomato worldwide [7, 30, 31]. However, the distribution and frequency of the species depend on the step of the tomato food chain, with ones more likely to be found on the field and others in greenhouses [32], during processing, or storage [33]. Tomatoes for the processing industry are produced under field conditions [7], while tomatoes for fresh consumption may be produced in both field and greenhouse conditions.

Fusarium is one of the most important fungal genera affecting the main crops worldwide, including several species with capability to cause diseases and mycotoxins contamination, which constitute a high risk of yield and quality losses [16]. In tomato, *Fusarium oxysporum* is one example, responsible for yield losses of 10 to 80%, depending on the region [7]. Specifically, *F. oxysporum f. sp. lycopersici* (*Fol*), which is the *forma specialis* affecting tomato plants [34], is responsible for about 14% of the economic losses in tomato crops, being the fifth most important plant pathogenic fungus worldwide [35]. Beyond its pathogenicity, *F. oxysporum* is a great producer of MON [25], also producing beauvericin (BEA), enniatins (ENNs), FUM, fusaric acid, and fusarins [36]. Other species and *formae speciales* of *Fusarium* genus also cause diseases on tomato, such as *F. equiseti* and *F. oxysporum* f. sp. *radicis-lycopersici* (*Forl*), that cause "*Fusarium* crown and root rot" [37, 38], *F. falciforme* and *F. solani* f. sp. *eumartii*, responsible for causing "*Fusarium* foot rot," and *F. striatum*, which causes "*Fusarium* crown and stem rot" [6]. From tomato fruits, the main *Fusarium* species isolated are *F. oxysporum, F. solani, F. equiseti* [12, 39], *F. semitectum, F. subglutinans* [39], *F. proliferatum,* and *F. verticillioides* [12], causing "*Fusarium* fruit rot" and potentially contaminating them with a panoply of mycotoxins [12, 36].

Nowadays, tomatoes are eaten fresh or processed into a great variety of products, such as sauces, ketchup, pulp, aseptic paste, juices, and canned and dried tomatoes [7, 23, 30], which may also be used as ingredients in a panoply of other foods [30]. This means that most of the population is often in contact with tomatoes and with the mycotoxins eventually present in them [33].

Despite mycotoxins contamination in fresh tomatoes is not usual, it is a real concern after processing, mostly due to the presence of *Alternaria* spp. that typically occurs under specific storage or processing environmental conditions that promote fungal growth, being temperature and humidity the major environmental factors [23, 40, 41]. The large consumption of tomatoes associated to the possible presence of mycotoxins may have a significant impact on human health [33]. In Europe, concerns were raised about the dietary exposure to *Alternaria* toxins that are chemically stable during food processing [41, 42]. In fact, there are reports of most of the tomato products contaminated with at least one *Alternaria* mycotoxin, whereas no *Alternaria* toxins were found in fresh tomatoes [43]. Recently, also *Fusarium* mycotoxins have been detected in tomatoes and derived tomato products, such as Fumonisin B2, one of those with higher values [23]. Other emergent *Fusarium* mycotoxins are ENNs A, A1, B and B1, and BEA, which are not regulated by governments worldwide [44, 45]. These emerging mycotoxins were detected in 58.8% of the tomato product samples (tomato powders, canned tomato, tomato sauce, ketchup, tomato juices, and dried tomatoes), but, curiously, not in fresh tomatoes [44, 45]. Thus, currently, there is a growing concern about the presence of mycotoxigenic fungi in tomato plants and fruits, being the control of fungi and their mycotoxins a crucial step to prevent yield losses and human health issues.

For that reason, in the last decades, some different ways of fungal and mycotoxins control have been developed, including some based on the use of biological control agents, such as antagonistic fungi, *Bacillus* and *Streptomyces* spp., endophytic bacteria and fungi, plant extracts, among others [46–52]. In fact, curiously, the concern with pests and diseases started long time ago in thethirteenth century, with Mexicans using agricultural practices as rotation and co-cultivation with different species to avoid the problems associated with high levels of moisture and certain pests [2]. For instance, red tomato was cultivated in the dry season due to its sensitivity to moisture, and husk tomato was co-cultivated with chili to decrease the incidence of pests [2].

Currently, one wide biocontrol measure against pathogens is the use of endophytic fungi, which reside in symbiosis within the plant and that is antagonist to the pathogen, disrupting its life cycle, preventing the colonization and infection, reducing the pathogen sporulation, and affecting its ability to survive [14]. However, also the endophytic fungi may be able to produce mycotoxins and to introduce them to the agri-food chain, posing a risk to the consumers [53]. For example, *Fusarium* is one of the endophytic fungal genera with significant potential to be used as biological control agent, comprising five or more known antagonistic species [54], such as *F. solani* strain K, which is known by increasing tomato defense against red spider mite infestations [55]. However, as previously mentioned, *Fusarium* is also a fungal genus that comprises relevant mycotoxigenic species. Therefore, to minimize the risk of mycotoxin contamination in tomatoes, following proper agricultural and postharvest practices is important.

This chapter aims to present the most relevant *Fusarium* species affecting tomato crops and their associated mycotoxins, as well as the consequences of consuming fresh tomato or industrially processed contaminated with mycotoxins and some promising biocontrol measures.

2. *Fusarium* diseases of tomato

The *Fusarium* genus is a very diverse group of filamentous fungi naturally present in soil and plants. It encompasses numerous species, with diverse roles in the ecosystems, since saprophytic or endophytic to phytopathogenic or mycotoxigenic ones. Moreover, some species can also cause infections in animals and humans, especially in the immunocompromised individuals [16], and most of them produce interesting metabolites with diverse applications. Its singularity and diversity turn *Fusarium* into one of the most relevant fungal genera in agriculture. Thus, this section is dedicated to the different tomato crop diseases caused by *Fusarium* species, their symptoms, distribution, and impact on production and economic yield.

2.1 "*Fusarium* wilt"

"*Fusarium* wilt" (FW) of tomato is one of the most frequent and destructive vascular disease responsible for extensive losses in the tomato crop production, both in the field and inside greenhouses, worldwide. This disease typically reduces the crop yield to minimal or absent [56]. However, yield losses are variable, with reports of 45% due to *Fol* in India and Cameroon [56, 57], or of 60–70% in Mexico [58].

The process of infection starts with the recognition of the roots by the fungi, attachment to the root surface, and penetration and colonization of the root cortex [34, 35]. The hyphae proliferate within the xylem vessels affecting the vascular tissue of the plant, clogging them with mycelium, spores, and/or polysaccharides produced by the fungus, which leads to the wilting symptoms, close to those of draught [34, 35, 59]. Furthermore, the fungus secretes several toxins in the vessels, such as fusaric acid, dehydrofusaric acid, lycomarasmin, and others, which, when carried to the leaves, cause reduced chlorophyll synthesis leading to reduced photosynthesis and decontrolled water loss through transpiration, also resulting in leaf wilting, interveinal necrosis, browning, and eventually death of the plant [59]. Therefore, the disease typically starts with wilting and yellowing of the lower leaves. The leaves or the infected parts of the plant loose turgidity and become gradually light green, yellow, brown, and finally die [35, 59]. As the fungus progresses through the plant, the wilt also spreads to the entire plant, and all leaves turn brown and become necrotic up to the whole plant death [59].

FW in tomato is mainly caused by species and *formae speciales* of the *F. oxysporum* complex. This species complex is worldwide spread and includes both nonpathogenic species, as well as species with the ability to cause disease in most of the crops, animals, and/or humans [16, 35, 60]. There are around 106 phytopathogenic *formae speciales* of *F. oxysporum* well documented affecting 45 plant families, being *Fol* the main responsible for FW in tomato [61]. *F. solani* can also be found in tomato plants causing FW, but mostly in varieties with resistance to *F. oxysporum* [62].

Fol was first described by Massee in 1985 in England [63] and was already documented in more than 30 countries [12, 58, 64], in several different environments and climates, from the dry and hot tropics up to the temperate climates [65], being the high temperatures and sandy and/or acidic soils the preferred conditions [66]. Currently, according to the tomato plant resistance genes that it overcomes, *Fol* comprises already three pathogenic races (race 1, race 2, race 3) [61], being expected a new race (race 4) until the end of 2023, considering the time observed between the appearance of two consecutive races [64]. Race 1 was described in 1886 and race 2 in 1945, in Ohio, having a worldwide distribution nowadays, while race 3 was firstly observed in Australia in 1978 and is present in a more limited region [35, 65]. Today, most of the commercial tomato varieties are resistant to races 1 and 2, but only few are resistant to race 3 [35], which results in higher yield losses.

Across the United States, there are reports of all three known physiological races of *Fol*, with the race 3 causing up to 80% of the disease incidence in tomato varieties resistant to races 1 and 2, whereas, in China, *Fol* race 1 is the mostly present but race 2, only reported in Zhejiang causes yield losses between 30 to 60% [7]. In Italy, FW is also mostly caused by *Fol* race 1 and race 2 and considered a major cause of economic loss in tomato production [7]. Also in the Southern Hemisphere, FW caused by *Fol* is a concern, being widely present in Brazil and Australia, and an emerging pathogen in Chile, causing huge yield and economic losses [7].

2.2 "*Fusarium* crown and root rot"

"*Fusarium* crown and root rot" (FCRR) is also a major fungal disease of both field and greenhouse tomato production [7].

FCRR is characterized by early loss of cotyledons, newest leaves, and basal stem necrosis [37]. Symptoms include wilting, where the leaves and stems become limp and lose their turgidity, yellowing of the leaves, due to a reduced ability to take up nutrients and water, stunted growth, with reduced overall plant size and vigor, because of the roots start to be darkening, browning, and rotting, losing the ability to support the plant's growth, while the crown shows lesions as dark and sunken areas on the stem at the soil line [37, 67]. The disease incidence and symptom severity increase with the inoculum concentration resulting, in the most severe cases, in plant collapse and death [67].

This disease is caused by both *Forl*, first reported in Japan in 1974 [7, 37], and by *F. equiseti* [38]. *Forl* is the main responsible for causing FCRR in tomato, but, contrary to *Fol*, is reported as a pathogen also in other plant hosts, such as eggplants, *Fabaceae* spp., and *Cucurbitaceae* spp. [61], and has no races identified [37, 61]. However, there are ten vegetative compatibility group (0090–0099) identified for *Forl* isolates from Israel, Belgium, Canada, Greece, France, Italy, Japan, and the United States, which can be divided in different subgroups [37]. *F. equiseti* is primarily a saprophyte or secondary a root colonizer, cosmopolitan in areas ranging from cool and temperate to hot and arid, which can affect a large range of plant hosts and even be an allergen or pathogen to humans [68]. Contrary to *F. oxysporum*, *F. equiseti* has not *formae speciales* reported.

FCRR causes losses estimated at up to 90% and 95% in Tunisian and Canadian greenhouses, respectively [69] and is present in all the USA tomato fields [37], causing a tomato yield reduction of 15–65% in Florida [70]. Also, in Italy, this disease causes substantial yield loss in both greenhouse and soil-less production, with a up to 60% mortality rate during severe *Forl* outbreaks [7] and being present in 24% of the greenhouses of Sicily and 66% of greenhouses in Sardinia during the 1990s [7]. *Forl* is also becoming an increasing threat in Brazil, with disease indices ranging from 10 to 50%, and Chile, affecting fresh tomato production more severely [7].

In Australia, *Forl* was not reported yet, but *F. oxysporum* strains were collected producing symptoms of rot on tomato roots and crowns such as those caused by *Forl*, but with different growth temperature, a larger host range, and variable pathogenicity, being this disease named as "chocolate streak disease" (CSD), different from FCRR caused by *Forl* [7].

2.3 "*Fusarium* foot rot"

"*Fusarium* foot rot" is a disease that is more prevalent in tomato crops grown during the cooler months and that usually occurs in patches in low-lying areas of the field [62]. Despite this disease does not often kill the tomato plants, it can substantially reduce the yield, up to 80% [62].

The fungus enters the cortical tissues of roots and hypocotyl by the existing wounds and causes brown lesions that extend into the vascular system, up to 25 cm of the root-stem transition zone. Despite the fungus kill the young seedlings, the infected plant usually does not wilt until its first full fruiting condition, and after fruit harvesting, or if atmospheric and other conditions reduce transpiration, the symptoms can reduce [71]. The symptoms may include varying degrees of chlorosis, mottling, and necrotic spotting on young foliage, and in severely affected plants, the main root and the crown can also be rotted [62, 72].

"*Fusarium* foot rot" in tomato can be caused by *Forl* [71], or by species and *formae speciales* from *F. solani* complex (FSSC), namely *F. falciforme* [72] and *F. solani f. sp. eumartii* [62]. *F. falciforme* was erstwhile called *Acremonium falciforme*, but, based on the results of more recent morphological and molecular phylogenetic analyses, it belongs to the *F. solani* complex now [73]. This species complex is easily isolated from soils in a variety of environments, comprising pathogenic species of a large number of plant species, including other economically important crops and trees [68]. *F. solani f. sp. eumartii* was firstly reported by Carpenter in 1915 in Pennsylvania, causing "Eumartii wilt" in potato [74]. However, this *forma speciales* of *F. solani* can also cause foot rot, tuber dry rot tubers and postharvest diseases of potato, foot rot of tomato, and stem and collar rot of eggplant and sweet pepper [74]. In the same way, *F. falciforme* can cause root rot in melon [75], lima bean [76], onion basal rot [77], root and stem rot in papaya [78], fruit rot of watermelon [79], and foot rot in tomato [72]. Furthermore, the *FSSC* comprises some of the most important *Fusarium* agents of animal and/or human infections in immunocompromised individuals [16, 68, 80]. For example, *F. falciforme* caused vertebral abscess and osteomyelitis in a 53-year-old woman with an autoimmune disorder [81].

2.4 "Fusarium crown and stem rot"

"*Fusarium* crown and stem rot" is a recent disease of the tomato crop, although it appeared in 2006 in commercial greenhouses of tomato plants in Canada and USA. Initially, pathologists were not able to identify the causal agent species, only knowing that it belonged to the FSSC. Only in 2014, the *Fusarium* isolates were identified as *F. striatum*, based on morphology and molecular analysis [6]. Recently, in 2020, it was also reported the presence of this disease since 2015–2016 in greenhouses in two states of Mexico [82].

"*Fusarium* crown and stem rot" is characterized by the development of dark brown cankers on stems and brown discoloration of the pith area. Symptomatic plants have brown lesions at the graph connections and at the pruning sites that evolve as cankers and lids to the dead of the plants within few weeks (8 weeks) after the appearance of the symptoms. The cankers are flat and depressed with a clear delineation from the stem. Plants with stem rot show chlorotic and senescent wilted leaves, and the whole plant exhibits wilting [6, 82]. In the base of diseased plants, the development of orange-red to dark-red perithecia on necrotic tissues and on rockwool slabs usually occurs. Plants can survive longer if the canker is removed within 2 weeks after the first symptoms were observed. Moreover, this new disease caused reminiscent of infection by *Forl*. A very typical symptom that allows to distinguish this disease from that caused by *Forl* is the discoloration of the pith rather than the vascular tissues that remain untouched [6]. Another distinguishing feature, but for the disease caused by *F. solani* f. sp. *Eumartii,* appears to be the homothallism of *F. striatum* in contrast to the heterothallism of *F. solani* f. sp. *Eumartii* [6, 83].

Several species for the FSSC are saprophytic, but there are also some that are phytopathogenic, causing severe losses mainly on horticultural crops. Within six years, from 2006 to 2012, *F. striatum* infected plants increased up to 20% in North America [6]. In greenhouses in Mexico, as the fungal infection is quite recent, the incidence was only up to 5% [82]. *F. striatum* was first described by Sherbakoff in 1915 who isolate the fungi from potato tubers originated in Colorado. A distinct characteristic is the formation of a "Striate rot" when inoculated in potatoes [6].

F. striatum in tomato was only reported twice, but it has other hosts, such as *Panax notoginseng,* where the first appearance was reported in 2020 in China [84]. In ginseng, *F. striatum* is an endophytic fungus with the ability to cause root rot disease. The isolate obtained from embryos could cause root rot in plants, which is a very interesting case of possible interchangeable functions, either as endophyte or as a pathogenic fungus. This situation poses difficulties during cultivation, where the fungus cannot be effectively prevented because it is a seed-borne pathogen. So far, such behavior has not been reported in tomato, but further studies should be conducted to understand the function of *F. striatum* in tomato seeds as the reported presence of the fungus in tomato plants was always in greenhouses.

2.5 *Fusarium* tomato fruit rot

"Tomato fruit rots" are important postharvesting diseases that compromise the quality of tomatoes and their possibility of being commercialized or consumed, resulting in economic losses and waste, that can reach over than 50% of the product during severe infections: 25% at harvest and 34% during transportation, storage, and market shops [85]. There are two types of tomato fruit rots: soft rots, such as "*Fusarium* tomato fruit rot" and dry rots. Soft rots of tomato fruit can be caused by *Erwinia carotovora, Rhizopus oryzae, R. stolonifera*, and *Fusarium* spp. and contribute to 85% of the overall loss of tomato [85].

"*Fusarium* tomato fruit rot" is characterized by soft rots that can extend into the center of the fruit. Usually, while the infected tissue is discolored and pale brown, the rotted tissue becomes covered by white, yellow, or pinkish mycelium [39]. Improperly handling, packaging, storage, and transportation of tomatoes promote the penetration of the fungus into the epicarp and their deposit in the fleshy parts of the mesocarp. Then, the fungus uses the ascorbic acid and soluble sugar contents in the fruit and proliferate, resulting in an enlarging necrosis of the epicarp toward the outside and in a more and more water-soaked fruit. The fungus continues extending deep into the seed cavity, destroys the seed coat and cotyledons, and eventually, the fruit completely rots [39].

The main species isolated from tomatoes with "*Fusarium* tomato fruit rot" are *F. solani* [12, 39], *F. oxysporum, F. equiseti* [12, 39, 85], *F. semitectum, F. subglutinans* [39], *F. proliferatum,* and *F. verticillioides* [12]. As previously mentioned, *F. oxysporum*, *F. solani,* and *F. equiseti* are also associated with other tomato diseases, but usually it is not the case of *F. semitectum, F. subglutinans* [39], *F. proliferatum,* and *F. verticillioides.* These species are mainly found affecting other crops, *for example,* maize or sorghum, being not considered common tomato pathogens, affecting mainly the postharvested tomato fruits. Their distribution is wide, and they can be found in different agricultural and nonagricultural substrates. For example, *F. semisectum* is commonly found in both soil and plants, not confined to just subtropical and tropical regions but also present in the Arctic and arid deserts. Furthermore, it is often associated with the development of storage rot in various types of fruits, including bananas and mushrooms. Additionally, all these four species can be allergenic to humans and cause severe infections with reduced treatment options due their resistance to most of the clinical antifungals [68]. *Fusarium solani*, *F. semitectum*, and *F. oxysporum,* including *Fol,* are the most abundant species isolated from tomatoes with "*Fusarium* fruit rot", while *F. verticillioides, F. proliferatum*, *F. subglutinans,* and *F. equiseti* represent a minor [12, 39].

3. *Fusarium* mycotoxins on tomato, their relevance, and control

3.1 *Fusarium* mycotoxins on tomato

Fresh tomatoes are rarely contaminated with mycotoxins, but their storage, even at low temperature, and processing may promote fungal growth and contamination with mycotoxins [43, 86]. As tomatoes have high water and nutrient contents, they can easily be contaminated with fungal species from *Aspergillus, Alternaria, Fusarium,* and *Penicillium* genera [45]. Thus, aflatoxins and ochratoxins produced mainly by *Aspergillus* spp., and FUM and fusaric acid, mainly produced by *Fusarium* spp., are mycotoxins that can potentially appear in tomatoes and their products. For example, aflatoxins were detected in rotten tomatoes commercialized in Nigeria [87]. Beyond these, some other emerging mycotoxins have been reported in tomato products, including alternariol monomethyl ether (AME), alternariol (AOH), and TeA, produced by *Alternaria* spp. [43], and BEA, ENNs, and MON produced by a wide variety of *Fusarium* spp. For example, TeA appears to be omnipresent in tomato products [43], and diverse assays report one or more of these emerging mycotoxins present in most of the tomato products analyzed [43, 45]. Therefore, frequent contamination of food and feed, including in tomatoes and tomato products, with emerging *Fusarium* and *Alternaria* mycotoxins has increased the interest about them [43]. These mycotoxins can provoke a wide range of toxicological effects to protect the plants, acting as bactericide, fungicide, antihelmintic, and insecticide, but they also may provoke phytotoxicity and cytotoxicity through mitochondrial modifications [88], decrease of the functionality of photosystems I and II leading to the reduction of photosynthetic pigments in the tomato leaves and to leaf wilting, cell dead, and necrosis [59].

As previously mentioned, the *Fusarium* species more frequently find in tomato plants are *F. oxysporum, F. equiseti, F. solani, F. falciforme,* and *F. striatum,* while in tomato fruits are *F. oxysporum, F. solani, F. equiseti, F. semitectum, F. subglutinans, F. proliferatum,* and *F. verticillioides.* However, if fresh tomatoes rarely show mycotoxin contamination because it occurs mostly during storage and processing, the main responsible species by mycotoxin contamination should be the ones found in tomato fruits. **Table 1** summarizes the mycotoxins produced by these species, and it is possible to see that all the most common species found in tomatoes produce BEA and/or MON, which justifies the emergence of these mycotoxins. Moreover, *F. oxysporum*, which also infects the tomato plants, and *F. proliferatum* are producers of ENNs, being a reason for its emergence in tomatoes and tomato products.

3.2 *Fusarium* mycotoxins relevance in human health

The large consumption of tomato products associated to the emergence of mycotoxins in them may have a significant impact on human health [33], resulting in acute or chronic consequences such as carcinogenic, teratogenic, immunosuppressive, or estrogenic issues [21]. In this way, the most common and known as harmful mycotoxins are regulated by the European Commission that set the maximum levels allowed

Fusarium species	Mycotoxins produced	References
F. oxysporum	BEA, ENNs, MON, FUM, fusaric acid, fusarins, bikaverin, and isoverrucarol	[25, 36, 68]
F. solani	MON, fusaric acid, and fusalanipyrone	[36, 68]
F. equiseti	BEA, MON, butenolide, equisetin, fusarochromanone, ZEN, trichothecenes, diacetoxyscirpenol, nivalenol, and T-2 toxin	[36, 39, 68, 89]
F. semitectum	BEA, MON, trichothecenes, ZEN, apicidins, equisetin, fusapyrone, sambutoxin, and diacetoxyscirpenol	[36, 39, 68, 89]
F. subglutinans	BEA, MON, fusaproliferin, fusaric acid, and FUM	[36, 39, 68]
F. proliferatum	BEA, ENNs, MON, fusaproliferin, fusaric acid, fusarins, and FUM	[36, 68]
F. verticillioides	BEA, FUM, fusaric acid, and fusarins	[36]

Table 1.
Summary of mycotoxins produced by the Fusarium species reported in tomato. Fruits.

of each individual mycotoxin [90]. These include aflatoxins, FUM, OTA, DON, ZEN, and ergot alkaloids [21, 22]. These limits turn all the products containing high levels of mycotoxins strongly devalued or even banned from being sold. However, the *in vivo* toxicity and toxicokinetic data about the three emergent mycotoxins in tomato (MON, BEA, and ENNs) are limited, as well as the data of their occurrence in food [43], which means that these mycotoxins are not regulated by governments worldwide, yet [45].

In vitro studies suggest that ENNs A, A1, and B1, BEA, and MON have genotoxic and immunomodulating effects, BEA and ENN B can cause a reproductive health hazard, and MON can cause severe toxicity in animals, mainly affecting the heart [43]. This might be a health concern, especially among children and the populations consuming a high amount of these products [45].

On the other hand, FUM, DON, and ZEN, also mainly produced by *Fusarium* species, are known as harmful to human health [16]. For example, FUM B1 is genotoxic and a cancer promoter, while ZEN is an estrogenic mycotoxin, being linked with endometrial adenocarcinoma and hyperplasia [21, 91]. Also, in an essay with rats, DON-induced DNA damage and the expression of proteins related to inflammatory response implied the apoptotic pathway [92].

As previously mentioned, one or more mycotoxins have been found in most of the tomato products analyzed in diverse essays. This multi-mycotoxin co-occurrence is, in fact, a frequent issue that may result in synergistic, additive, or antagonistic effects of mycotoxins. For example, when aflatoxins are present together with DON and T2 toxin, OTA with FUM, or FUM with DON, there is a synergetic effect of these mycotoxins [21, 93, 94]. Thus, it is urgent to consider the toxicological effects of mycotoxins mixtures in the new risk assessments [94].

Considering, the emergence of mycotoxins in tomato products and the lack of knowledge about their real effects on human health; currently, there is a growing concern about the presence of mycotoxigenic fungi in tomato plants, fruits, and products, being the control of these fungi and their mycotoxins a crucial step to prevent both yield losses and human health issues.

3.3 Mycotoxins control measures

All stages of production, harvesting, transport, storage, and processing of tomatoes might undergo to mycotoxin contamination [45]. However, since mycotoxins are not usually found in fresh tomatoes, the storage and processing of tomatoes appear to be the most critical stages.

Nonetheless, within the food supply chain, similar to the feed supply chain as both are integrated, mycotoxin management must be comprehensive, encompassing all phases from the cultivation of raw materials until the production of food [95]. During tomatoes production, good farming practices should be applied to minimize the mycotoxigenic fungi presence. Postharvesting during transportation, storage, processing, sanitation and controlling humidity, temperature, ventilation, insects, and rodents presence are also good practices to avoid fungal proliferation and mycotoxins production [95, 96]. It is also important to keep in mind that most of the mycotoxins are chemically stable to heat during processing [45].

There are also new strategies that may be used, inclusive in combination, such as using bio-competitive fungi as biocontrol agents, varieties resistant to pathogens, apply antibody-mediated technology, or even nanomaterials engaged for antifungal or inhibition of mycotoxins [95]. Also, some physical, thermal, and biological methods may be used such as cleaning and aggressive sorting, nonionizing and ionizing irradiation, or microorganisms with detoxification activities [95, 96]. Other methods to control fungal and mycotoxin contamination have been studied, such as ozone application, that inhibits fungal growth and cold plasma technology, which reduces and degrades mycotoxins in food and feed, pulsed light, radio frequency, and microwave for reducing mycotoxin contamination [95]. Additionally, some binding agents (inorganic or organic) or bio-transforming agents (*e.g.*, bacteria, yeast, fungi, and enzymes that degrade mycotoxin molecules in nontoxic metabolites) may be used to control fungal proliferation and mycotoxin contamination [95, 96].

Nevertheless, despite the availability of various strategies to manage mycotoxins by either pre-harvest prevent or post-harvest eliminate/decontaminate, achieving their total eradication remains very difficult. The precision of detection methods plays a fundamental role on the control of contaminations [45, 95].

4. Management of diseases caused by *Fusarium* spp.

Some good farming practices can help reducing the risk of disease development and minimize the severity of fungal diseases. However, it is important to note that managing *Fusarium* diseases can be challenging as the pathogen can persist as chlamydospores in the soil or as mycelium in infected plant debris for long periods [97]. Thus, a combination of practices, including prevention, sanitation, single-gene resistance, and cultural management is usually the most effective approach to control the disease and minimizing the impact on tomato crop.

Prevention may be accomplished firstly by choosing an adequate planting site. This should be characterized by a well-drained, not sandy or acidic soil, without high temperatures or high fluctuations between day and night, and with good air circulation [95, 96]. For instance, *Fol* is favored by high temperatures in the soil, but *F. solani* may be favored by a temperature of 18°C or 28°C depending on high or low soil moisture, respectively [98].

Secondly, it is recommended to utilize effective irrigation systems such as drip irrigation. This approach prevents leaf sprinkling and excessive watering [30], which could potentially disperse the pathogen over greater distances [66]. Also, giving proper spacing between contiguous plants and doing a proper fertilization avoiding excessive nitrogen are wise [95, 96].

Seed sanitation by means of several different treatments as heat or thermotherapy may decrease seed-borne infections. In fact, efforts have been undertaken to manage *Fusarium* spp. by regulating not only the quality of propagating plant material like seed and seedlings, but also the quality of planting substrates, trays and other items used in nurseries and greenhouses [99].

Furthermore, implementing an integrated pest management strategy and choosing tomato varieties that exhibit resistance to specific *Fusarium* species and races represent straightforward, cost-effective, and efficient preventive measures [58]. In fact, plant resistance to pathogens is theoretically the most economical, effective and environmentally friendly approach to disease management [37, 100]. However, no variety was so far completely resistant to, for instance, *Fol*. Akram and co-workers (2014) [97] verified that only three varieties were moderately resistant to *Fol,* being all the others either susceptible or very susceptible. Other authors reported several varieties resistant either to all races of *Fol* and susceptible to *Forl* or resistant to combinations of two races of *Fol* and to *Forl* [37]. For each of the three *Fol* races, resistance has been introgressed from wild tomato species, using four R genes [100]. The relationship host-pathogen is under constant modification/adaptation with the pathogens gradually overcoming the tomato cultivar resistance mechanisms. Plant breeding had been limited success due to the emergence of new pathogenic races [101], combined with the lack of knowledge on the mechanisms underlying the host-pathogen interaction, which may evolve to infection or to the resistance of the host [100].

Doing a good soil management is also important, namely doing crop rotation with non-susceptible plants, which can help reducing the fungal population in the soil that is pathogenic to tomato crops [95, 96], or even opting for soil-less systems, which offer a unique opportunity to use antagonists. Biological control of *Fusarium* diseases by means of action of nonpathogenic strains of *F. oxysporum* may be easier with its application in soil-less cultures of tomatoes than on soil cultures, considering the need of high mass-production of the antagonist and its incorporation in an appropriate formulation to be applied on several hectares [47]. Investing in sanitation is also important to limit the disease's spread and can be done, for example, removing and destroying infected plants and choosing seeds with high sanitary quality, since seeds may be vectors of the fungi at long distances [66].

Employing fungicides is a widespread practice, but it frequently fails to meet the expected levels of effectiveness due to the presence of fungi in both the soil and seeds, which poses a challenge [97]. *Fol* has been controlled by disinfection of the soil with chemicals as methyl bromide with damaging effects on the environment and on human and animal health [102]. The extensive use of fungicides, beside having negative impact in the environment, promotes the emergence of resistant strains and may kill the endophyte fungi that collaborate with the plant and ensure its virility [16]. Furthermore, this measure needs to be applied several times and in the proper time to be effective as a preventive method, which implies significative economic costs [101, 103]. To prevent and minimize these negative impacts, research on new and biological alternatives to chemical fungicides is a rising issue, including the use of bacteria or other fungal species [24, 54] for biocontrol of harmful species. The simultaneous use of biocontrol agents and synthetic fungicides [54], as well as genetic engineering, *for example,* introducing genes, which encode proteins with antifungal activity into the plant genome, are alternative approaches that can be used in some cases [100, 104].

4.1 Biocontrol measures

Biological control has received increasing attention and acceptance in the last decades. It aims to control pests and diseases importing, reinforcing, or conserving other organisms (biocontrol agents) in the environment [7]. Moreover, its high efficiency, target-specificity, and self-sustainability minimize the public concerns of impacts on health and environment, resulting in a higher public acceptance and advantages compared to the conventional methods [7].

To biocontrol fungal plant diseases, a mix of different strategies can be applied. Antagonist species can be used to produce inhibitory metabolites, act as parasites on pathogenic fungi, or compete with pathogenic species [7, 54, 105]. Some of the most used biocontrol agents are *Aspergillus* spp., *Chaetomium* spp., *Glomus* spp., nonpathogenic *Fusarium* spp., *Trichoderma* spp., *Penicillium* spp., *Bacillus* spp., *Pseudomonas* spp., *Streptomyces* spp., or *Serratia* spp. [106]. Biocontrol of *Fusarium* spp. associated diseases is a sustainable and environmentally friendly approach to mitigate the impact of the pathogens on the tomato crop, on the ecosystem, and on the consumers. This approach is an alternative to conventional fungicides and fumigation by methyl bromide, which are known by their negative effects on the human health, animal well-being, and environment (e.g., depletion of the ozone layer) [107].

For example, nonpathogenic *Fusarium* species and fluorescent *Pseudomonas* are reported as the more effective and consistent biocontrol agents against FW in tomato produced in natural soil, while strains from *Penicillium*, *Streptomyces*, and *Aspergillus* species are mostly effective *in vitro* conditions [7].

Also, metabolites produced by these and other organisms may induce the plant defense response against pathogens, reducing the effects of disease [7, 54, 105].

In tomato, there are reports of bacteria, fungi, and plant and algae extracts effectively used as biocontrol agents against diseases caused by *Fusarium* spp. [7]. In fact, some *Fusarium* strains affecting tomato are known as having poor competitive fitness against other microorganisms, such as *Forl*, being the biocontrol using antagonistic organisms effective. Thus, several antagonistic microorganisms have been successfully tested against *Forl* and *Fol* [7]. However, around 79% of the tests on tomatoes were conducted in greenhouse conditions, and only 12% in the field condition, which means that further field tests are necessary [7].

4.1.1 Bacteria as biocontrol agents against Fusarium *diseases in tomato*

Some soilborne bacteria, such as *Pseudomonas, Bacillus, or Streptomyces spp., for example,* can be used as competitors against *Fusarium* species.

Some of the most relevant *Pseudomonas* species used as biocontrol agents in tomato are *P. fluorescens*, *P. chlororaphis*, and *P. putida*. For example, *P. chlororaphis* can reduce the incidence of FCRR in 44–60%, and *P. putida* can reduce the incidence of FW in 41–94% [37]. *P. fluorescens* was also applied to control FW with success, what was concluded by the reduction of the incidence in 53–72% or 65–85%, depending on the production system, in greenhouse, or in field, respectively [37], and, *in vitro*, *P. fluorescens* can suppress the growth of *Forl* synthesizing the antibiotic 2,4-diacetylphloroglucinol [7]. *P. fluorescens* can also be used combined with the fungicide benomyl to decrease FW incidence more effectively in greenhouse environment [108].

From *Bacillus* genus, some of the most relevant species used as biocontrol agents in tomato are *B. subtilis, B. amyloliquefaciens, B. cereus*, and *B. pumilus* [109]. In India, *B. amyloliquefaciens* and *B. subtilis* were reported as capable to reduce the incidence of FW in 44–46% and 53–64%, respectively, under field conditions [37]. In vitro, *B. amyloliquefaciens* also can inhibit the growth of *F. solani* by 95.2% and *B. cereus* reduced the growth of the fungus in 55.7%. Also, *B. pumilus* and *B. subtilis* can inhibit *F. solani* growth in 70.46% and 82.1%, respectively [109]. The same biocontrol potential was verified *in vivo*, with *B. cereus* controlling 81.2% of the disease, *B. amyloliquefaciens* controlling 75%, and *B. pumilus* and *B. subtilis* controlling 62.5% individually [109]. Some strains of *B. pumilus* (*e.g.*, TolrMA) also can reduce FW incidence in about 73% and increase the root and shoot length 60% and 84%, respectively [49]. While some *B. subtilis* strains (*e.g.*, QST713 and EU07) can reduce the FCRR incidence by 52% and 75%, respectively [110].

Also, *B. subtillis* can be used combined with a fungicide, such as hymexazol or the resistance elicitor acibenzolar-s-methyl, to produce greater reductions in the FCRR disease incidence than any of these treatments alone [111].

Endophytic bacteria also can be used as biocontrol of FW and FCRR, including species from other plants neither *Solanum lycopersicum*. For example, essays using endophytic bacteria from *Solanum sodomaeum*, *Solanum bonariense*, and *Nicotiana glauca* reported their capability to control *Fol* [50, 51]. Also, endophytic *Streptomyces* spp. strains (*e.g.*, SNL2) isolated from roots of native plants can be effective against *Forl*, reducing the FCRR incidence in 75.3%, and significantly increasing the dry weight, shoot, and root length of tomato seedlings [48]. In the same way, Strept*omyces griseus* can reduce the severity of FW in 57% in greenhouse conditions [37]. Any plant rhizosphere may carry antagonistic bacteria that can suppress pathogens, as well as enhance tomato plant growth [112]. For example, some strains of *B. subtilis* (*e.g.*, BsTA16) and *Acinetobacter calcoaceticus* (*e.g.*, AcDB3) have potential against *Fol* race 3 or *Forl* in tomato, respectively [112]. Another example is *Achromobacter xylosoxidans* that can reduce the incidence of FW in 50% in greenhouse environments [37].

4.1.2 Fungi as biocontrol agents against Fusarium diseases in tomato

The fungal high reproductive rate, short generation time, and high capability to survive both in the environment, such as parasites or saprotrophs, and in the plant, such as endophytic, turn fungi an interesting, potential, and sustainable option to biological control [14, 54]. In one hand, these species can use competition for space and nutrition, volatile and diffusible antibiosis, and hydrolytic enzyme production to minimize the pathogen inoculum. In the other hand, they can also colonize the plant roots, to induce growth and nutrient adsorption for the plant, and invade the vascular tissue or epidermal cells of plant root, to induce systemic resistance in plant, and defend it from the pathogen infection [105]. For example, nonpathogenic strains of fungi can be used as competitors against *Fusarium* species causing FW, helping suppress this disease, like nonpathogenic isolates of *F. solani, or certain Trichoderma species, for example* [7, 68, 105].

Trichoderma is one of the greatest well-known fungal biocontrol agents against plant fungal diseases, comprising 25 biocontrol agents [54, 105]. In tomato crops, *Trichoderma asperellum* MSST is an efficient promising biocontrol agent against *Fol*, inhibiting its entrance in the plant root, inducing the systemic resistance in the plant, and reducing the FW incidence up to 85% [105]. Another *Trichoderma* species known for its effectiveness against *Fusarium* diseases is *T. harzianum*. When used as a seed coating or mixed with wheat-bran/peat for tomatoes cultivated in fields infested with *Forl*, it led to a remarkable 26.2% increase in yield [113]. Research carried out in Thailand, Israel, Egypt and the USA has demonstrated that the application of *T. harzianum* leads to a significant decrease in both the disease incidence and severity. The reduction in incidence ranged from 30% to 80%, while severity reduction varied between 24% and 44%. These variations are influenced by the specific region, although there is some overlap in the observed values [37]. *T. harzianum* can also be applied combined with *Rhizophagus intraradices,* with *Aspergillus ochraceus* and *Penicillium funiculosum*, or with *Pseudomonas* sp. and *R. intraradices*, reducing, even more, the severity and incidence of these diseases [37, 114].

Other *Aspergillus* and *Penicillium* species with biocontrol application are *A. nidulans*, *A. awamori*, and *P. digitatum*, which can reduce the severity of FW by 63%, 37%, and 21%, respectively [37].

Using endophytic fungi is also a wise measure of biocontrol since it resides on the natural symbiosis established within the plant between the antagonist and the pathogen. The antagonist disrupts the pathogen life cycle, prevents the colonization and infection, reducing its sporulation, and ability to survive, and therefore protecting the plant [14]. For example, nonpathogenic endophytic *F. solani* strains may reduce the incidence of diseases caused by *Forl* by 47%, or even more if combined with certain fungicides [115]. Also, nonpathogenic strains of *F. oxysporum* can induce the systemic resistance and defense reaction of the tomato plant and successfully control *Fol* [7] and *Forl*, reducing the severity and incidence of FW in 38–58% and 57–78%, depending on the production done in greenhouses or in field condition (respectively), and the incidence of FCRR in 78% in greenhouses [37].

Other fungi also used as biocontrol agents are *Chaetomium lucknowense* and *C. globsum*. Both can reduce the severity of FW in 36% and 44% [37]. Additionally, *Pythium oligandrum* can control *Forl* and FCRR in tomato [54]. Also, *Rhizoctonia* strains can reduce completely the vascular discoloration caused by *Forl* on tomatoes in greenhouse conditions and up to 70% in the field [116], and some mycorrhizae species, such as *R. intraradices*, can effectively reduce the severity of FW in GH (72%), as well as the incidence and severity of FCRR in field (18–71% and 16–53%, respectively) [37].

4.1.3 Other biological solutions against Fusarium diseases in tomato

Biopesticides, *that is*. natural products derived from living organisms, such as plants, extracts, and algae with fungicidal properties, are being studied and developed to control pests in agriculture and forestry with reduced reliance on synthetic chemicals [52, 117, 118].

Aqueous extracts from plants such as, for instance, garlic and the tropical plants *Callistemon citrinus, Cymbopogon citratus*, and *Oxalis barrelieri,* have anti-FW potential properties by inhibiting mycelial growth or by stimulating the tomato defense system [52, 117]. The disease incidence decreased in treated plants with percentages of reduction ranging from 51–84% depending on the treatment [52]. The authors concluded that these compounds will be helpful in the management of the disease caused by *Fol* even under field conditions [117] improving productivity and the quality of the production [52].

Macroalgae are also very promising organisms against wilt disease in tomato as they contain several bioactive antifungal complexes that can be used in the field [118].

Essential oils extracted from *Cymbopogon citratus* and *Ocimum gratissimum* were very active against radial growth and conidia germination of *F. oxysporum* with 100% inhibition of mycelial growth. The inhibitory effect is due to the presence of phenols, flavonoids, tannins, and coumarins [56]. Neem oil, which is derived from the neem tree, was the essential oil that had the strongest inhibition on *F. oxysporum* comparatively to garlic oil and argan oil due to the presence of azadirachtin, a terpenoid with antifungal activity, reaching 40.54% [119]. Linalool is also a common volatile monoterpene alcohol found in plants with a wide range of fungicide activity, namely against the soilborne *Forl*. The remaining doubts that still hamper its wide use in the field are the effects on soil-beneficial microorganisms [120].

All these findings are contributing for the development of "green" fungicides.

Organic amendments are another group of biocontrol agents that can be used against *Fusarium infections* in tomato. For example, the combination of pelletized poultry manure with heating or solarization can significantly reduce the FW severity [121]. Also, the combined application of biocontrol organisms and amendments can result in higher biocontrol efficiency of diverse genera of bacteria or fungi, excepting *Pseudomonas* and *Penicillium* [7].

Herbal vermicompost and vermicompost biofortified with selected biological control agents (such as *T. harzianum, P. fluorescens, and B. subtilis*) significantly increased soil pH, ammonium nitrogen, soil organic matter, and dissolved organic carbon, providing abundant antagonists and promoting beneficial bacteria that enhanced plant growth, yield, and nutritional quality, which, in turn, suppressed *Fol* effects [122].

Wang et al. (2021) [107] observed the effect of vermicompost on promoting growth of Actinobacteria, Chloroflexi, Saccharibacteria, and Planctomycetes and inhibiting growth of Proteobacteria, Gemmatimonadetes, Firmicutes, Verrucomicrobia, and Cyanobacteria. This alteration in the relative presence of Ascomycota and Basidiomycota reduced the incidence of *Fol* by 36.5% up to 73.9%, being the protective effect increasing in proportion to the rate of application [123]. In this way, vermicompost has still to be tested and exploited for large-scale field applications.

5. Conclusions

Within the tomato crop, the prevalent pathogenic *Fusarium* species include *Fol*, *Forl*, and *F. solani*. These particular species are responsible for inducing a range of diseases such as vascular wilts, root, stem and crown rots, and cankers, which have a global impact on both field-grown and greenhouse-cultivated tomatoes. *Fusarium* fungi survive as chlamydospores for long periods in the soil and wilt strains colonize the root cortex of some non-host plants, which behave as natural reservoirs of propagating structures. Thus, controlling *Fusarium* is challenging. Currently, single-gene editing and preferably dual-gene CRISPR/cas9 already demonstrated the ability to improve tolerance/resistance to fusarium wilt, but experimental field evaluation is still needed.

Fusarium species produce a range of mycotoxins, such as FUM, DON, and ZEN, which harm animal and human health. The wilt species, beside clogging the vessels, also can produce mycotoxins, such as fusaric acid, lycomarasmin, and dehydrofusaric acid, being fusaric acid a contaminant in tomatoes and the tomato-based products. Furthermore, most of the Fusarium spp. found in tomatoes can produce BEA, MON, and/or ENNs, which have been detected more frequently and of which effects in human health are unknown. Therefore, these fungal species have a great impact on the international economy, food safety, and security.

Further studies are needed to develop tools and models to early detect the diseases, to control the fungi, and to mitigate their effects. Under the current climatic change, a new challenge is related with the emergence of exotic new species and their associated mycotoxins in tomato and tomato-based products.

Acknowledgements

FCT - Fundação para a Ciência e Tecnologia, I.P. supported D.S. by the Ph.D. fellowship UI/BD/154444/2022. This work was also supported by FCT, Portuguese Foundation for Science and Technology through the R&D Unit, UIDB/04551/2020 (GREEN-IT, Bioresources for Sustainability).

Conflict of interest

The authors declare no conflict of interest.

DOI: http://dx.doi.org/10.5772/intechopen.1003643

Author details

Daniela Simões[1] and Eugénia de Andrade[2,3]*

1 Centre for Ecology, Evolution and Environmental Changes, Faculty of Sciences of University of Lisbon, Lisbon, Portugal

2 National Institute for Agricultural and Veterinary Research (INIAV), I.P., Oeiras, Portugal

3 GREEN-IT Bioresources for Sustainability, ITQB NOVA, Oeiras, Portugal

*Address all correspondence to: eugenia.andrade@iniav.pt

References

[1] Jenkins JA. The origin of the cultivated tomato. Economic Botany. 1948;**2**:379-392. DOI: 10.1007/BF02859492

[2] Saavedra TM, Figueroa GA, Cauih JGD. Origem e evolução da produção de tomate Lycopersicon esculentum no México. Ciencia Rural. 2017;**47**:e20160526. DOI: 10.1590/0103-8478cr20160526

[3] FAO Agricultural production statistics 2000-2021. FAOSTAT Anaytical Brief 60. 2022

[4] FAO Crops and Livestock Products. Available from: https://www.fao.org/faostat/en/#data/QCL/visualize [Accessed: September 8, 2023].

[5] FAO FAOSTAT – Tomatoes. Available from: http://data.un.org/Data.aspx?d=FAO&f=itemCode%3A388 [Accessed: September 8, 2023]

[6] Moine LM, Labbé C, Louis-Seize G, Seifert KA, Bélanger RR. Identification and detection of fusarium striatum as a new record of pathogen to greenhouse tomato in Northeastern America. Plant Disease. 2014;**98**:292-298. DOI: 10.1094/PDIS-08-13-0844-RE

[7] Ma M, Taylor PWJ, Chen D, Vaghefi N, He JZ. Major Soilborne pathogens of field processing tomatoes and management strategies. Microorganisms. 2023;**11**:1-29. DOI: 10.3390/microorganisms11020263

[8] Panthee D, Chen F. Genomics of fungal disease resistance in Tomato. Current Genomics. 2009;**11**:30-39. DOI: 10.2174/138920210790217927

[9] Gleason ML, Edmunds BA. Tomato diseases and disorders. Iowa State University Extension publication. PM 1266 revised August 2006. pp. 1-12

[10] Rodrigues MHP, Furlong EB. Fungal diseases and natural defense mechanisms of tomatoes (Solanum lycopersicum): A review. Physiological and Molecular Plant Pathology. 2022;**122**:101906. DOI: 10.1016/j.pmpp.2022.101906

[11] Pavan Kumar S, Srinivasulu A, Raja Babu K. Symptomology of major fungal diseases on tomato and its management. Journal of Pharmacognosy and Phytochemistry. 2018;7:1817-1821

[12] Murad NBA, Kusai NA, Zainudin NAIM. Identification and diversity of Fusarium species isolated from tomato fruits. Journal of Plant Protection Research. 2016;**56**:231-236. DOI: 10.1515/jppr-2016-0032

[13] Janowski D, Leski T. Factors in the distribution of Mycorrhizal and soil fungi. Diversity. 2022;**14**:1-15. DOI: 10.3390/d14121122

[14] Hyde KD, Xu J, Rapior S, Jeewon R, Lumyong S, Niego AGT, et al. The amazing potential of fungi: 50 ways we can exploit fungi industrially. Fungal Diversity. 2019;**97**:1-136. DOI: 10.1007/s13225-019-00430-9

[15] Naranjo-Ortiz MA, Gabaldón T. Fungal evolution: Major ecological adaptations and evolutionary transitions. Biological Reviews. 2019;**94**:1443-1476. DOI: 10.1111/brv.12510

[16] Simões D, de Andrade E, Sabino R. Fungi in a one health perspective. Encyclopedia. 2023;**3**:900-918. DOI: /10.3390/encyclopedia3030064

[17] Fones HN, Bebber DP, Chaloner TM, Kay WT, Steinberg G, Gurr SJ. Threats to global food security from emerging fungal and oomycete crop pathogens.

Nature Food. 2020;**1**:332-342. DOI: 10.1038/s43016-020-0075-0

[18] Peng Y, Li SJ, Yan J, Tang Y, Cheng JP, Gao AJ, et al. Research progress on Phytopathogenic fungi and their role as biocontrol agents. Frontiers in Microbiology. 2021;**12**:670135. DOI: 10.3389/fmicb.2021.670135

[19] Daley DK, Brown KJ, Badal S. Chapter 20 - Fungal Metabolites. In: Badal S, Delgoda R, editors. Pharmacognosy: Fundamentals, Applications and Strategy. Cambridge, MA, USA: Academic Press; 2017. pp. 413-421. ISBN 9780128020999

[20] Esheli M, Thissera B, El-Seedi HR, Rateb ME. Fungal metabolites in human health and diseases—An overview. Encyclopedia. 2022;**2**:1590-1601. DOI: 10.3390/encyclopedia2030108

[21] Afsah-Hejri L, Jinap S, Hajeb P, Radu S, Shakibazadeh S. A review on mycotoxins in food and feed: Malaysia case study. Comprehensive Reviews in Food Science and Food Safety. 2013;**12**:629-651. DOI: 10.1111/1541-4337.12029

[22] Seyedmousavi S, Bosco SDMG, De Hoog S, Ebel F, Elad D, Gomes RR, et al. Fungal infections in animals: A patchwork of different situations. Medical Mycology. 2018;**56**:165-187. DOI: 10.1093/mmy/myx104

[23] Van de Perre E, Deschuyffeleer N, Jacxsens L, Vekeman F, Van Der Hauwaert W, Asam S, et al. Screening of moulds and mycotoxins in tomatoes, bell peppers, onions, soft red fruits and derived tomato products. Food Control. 2014;**37**:165-170. DOI: 10.1016/j.foodcont.2013.09.034

[24] Ajilogba CF, Babalola OO. Integrated management strategies for tomato fusarium wilt. Biocontrol Science. 2013;**18**:117-127. DOI: 10.4265/bio.18.117

[25] Logrieco A, Bottalico A, Mulé G, Moretti A, Perrone G. Epidemiology of toxigenic fungi and their associated mycotoxins for some Mediterranean crops. European Journal of Plant Pathology. 2003;**109**:645-667. DOI: 10.1023/A:1026033021542

[26] Agrios GN. Botrytis diseases. In: Dreibelbis D, editor. Plant Pathology. San Diego, CA: Elsevier Academic Press; 2005. pp. 510-514 ISBN 0-12-473276-3

[27] Williamson B, Tudzynski B, Tudzynski P, Van Kan JAL. Botrytis cinerea: The cause of grey mould disease. Molecular Plant Pathology. 2007;**8**:561-580. DOI: 10.1111/j.1364-3703.2007.00417.x

[28] Hua L, Yong C, Zhanquan Z, Boqiang L, Guozheng Q, Shiping T. Pathogenic mechanisms and control strategies of Botrytis cinerea causing post-harvest decay in fruits and vegetables. Food Quality and Safety. 2018;**2**:111-119. DOI: 10.1093/fqsafe/fyy016

[29] Blum LEB. Reduction of incidence and severity of Septoria lycopersici leaf spot of tomato with bacteria and yeasts. Ciência Rural. 2000;**30**:761-765. DOI: /10.1590/S0103-84782000000500003

[30] Yücel S, Can C, Yurtmen M, Cetinkaya-Yildiz R, Aysan Y. Tomato pathology in Turkey. The European Journal of Plant Science and Biotechnology. 2008;**2**(1):38-47

[31] Ahmed FA, Sipes BS, Alvarez AM. Postharvest diseases of tomato and natural products for disease management. African Journal of Agricultural Research. 2017;**12**:684-691. DOI: 10.5897/ajar2017.12139

[32] Sanoubar R, Barbanti L. Fungal diseases on tomato plant under

greenhouse condition. European Journal of Biological Research. 2017;7:299-308. DOI: 10.5281/zenodo.1011161

[33] Mariutti LRB, Soares LMV. Survey of aflatoxins in tomato products. Ciência e Tecnologia de Alimentos. 2009;**29**:431-434. DOI: 10.1590/s0101-20612009000200031

[34] Michielse CB, Rep M. Pathogen profile update: Fusarium oxysporum. Molecular Plant Pathology. 2009;**10**:311-324. DOI: 10.1111/j.1364-3703.2009.00538.x

[35] Srinivas C, Nirmala Devi D, Narasimha Murthy K, Mohan CD, Lakshmeesha TR, Singh BP, et al. Fusarium oxysporum f. sp. lycopersici causal agent of vascular wilt disease of tomato: Biology to diversity– A review. Saudi. Journal of Biological Sciences. 2019;**26**:1315-1324. DOI: 10.1016/j.sjbs.2019.06.002

[36] Munkvold GP. Fusarium species and their associated mycotoxins. Methods in Molecular Biology. 2017;**1542**:51-106. DOI: 10.1007/978-1-4939-6707-0_4. PMID: 27924531

[37] McGovern RJ. Management of tomato diseases caused by Fusarium oxysporum. Crop Protection. 2015;**73**:78-92. DOI: 10.1016/j.cropro.2015.02.021

[38] Beltrán Beache M, Delgado Ortiz JC, Valdivia Flores AG, Hernández Juárez A, García Munguía AM. First report of Fusarium equiseti causing root and crown rot in tomato in Mexico. Plant Disease. 2023;**107**:2542. DOI: 10.1094/pdis-10-22-2494-pdn

[39] Tomato LTRB, Bakar AA, Izzati MNA, uMI KALSoM Y. Diversity of Fusarium species associated with post-harvest fruit rot disease of Tomato. Sains Malaysiana. 2013;**42**:911-920

[40] Hasan HAH. Alternaria mycotoxins in black rot lesion of tomato fruit: Conditions and regulation of their production. Acta Microbiologica et Immunologica Hungarica. 1996;**43**:125-133

[41] Puntscher H, Marko D, Warth B. The fate of altertoxin ii during tomato processing steps at a laboratory scale. Frontiers in Nutrition. 2019;**6**:1-6. DOI: 10.3389/fnut.2019.00092

[42] Arcella D, Eskola M, Gómez Ruiz JÁ. Dietary exposure assessment to Alternaria toxins in the European population. European Food Safety Authority Scientific Report. EFSA Journal. 2016;**14**:4654. DOI: 10.2903/j.efsa.2016.4654

[43] Fraeyman S, Croubels S, Devreese M, Antonissen G. Emerging fusarium and alternaria mycotoxins: Occurrence, toxicity and toxicokinetics. Toxins (Basel). 2017;**9**:1-26. DOI: 10.3390/toxins9070228

[44] Serrano AB, Font G, Mañes J, Ferrer E. Emerging Fusarium mycotoxins in organic and conventional pasta collected in Spain. Food and Chemical Toxicology. 2013;**51**:259-266. DOI: 10.1016/j.fct.2012.09.034

[45] Ji X, Deng T, Xiao Y, Jin C, Lyu W, Wu Z, et al. Emerging Alternaria and Fusarium mycotoxins in tomatoes and derived tomato products from the China market: Occurrence, methods of determination, and risk evaluation. Food Control. 2023;**145**:109464. DOI: /10.1016/j.foodcont.2022.109464

[46] Choi HW, Ahsan SM. Biocontrol activity of aspergillus terreus ANU-301 against two distinct plant diseases, Tomato Fusarium wilt and potato soft rot. Plant Pathology Journal. 2022;**38**:33-45. DOI: 10.5423/PPJ.OA.12.2021.0187

[47] Alabouvette C, Lemanceau P, Steinberg C, De Flore L, Faune D, Cedex D. Recent advances in the

biological control. Journal of Pesticide Science. 1993;**37**:365-373

[48] Goudjal Y, Zamoum M, Sabaou N, Mathieu F, Zitouni A. Potential of endophytic Streptomyces spp. for biocontrol of Fusarium root rot disease and growth promotion of tomato seedlings. Biocontrol Science and Technology. 2016;**26**:1691-1705. DOI: 10.1080/09583157.2016.1234584

[49] Heidarzadeh N, Baghaee-Ravari S. Application of Bacillus pumilus as a potential biocontrol agent of Fusarium wilt of tomato. Archives of Phytopathology and Plant Protection. 2015;**48**:841-849. DOI: 10.1080/03235408.2016.1140611

[50] Aydi-Ben-Abdallah R, Jabnoun-Khiareddine H, Daami-Remadi M. Fusarium wilt biocontrol and tomato growth stimulation, using endophytic bacteria naturally associated with Solanum sodomaeum and S. Bonariense plants. Egyptian Journal Of Biological Pest Control. 2020;**30**:1-13. DOI: 10.1186/s41938-020-00313-1

[51] Abdallah RAB, Mokni-Tlili S, Nefzi A, Jabnoun-KhiareddineH,Daami-RemadiM. Biocontrol of Fusarium wilt and growth promotion of tomato plants using endophytic bacteria isolated from Nicotiana glauca organs. Biological Control. 2016;**97**:80-88. DOI: 10.1016/j.biocontrol.2016.03.005

[52] Dakole Daboy,C, Azah Udom R, Fonkoua M, Dongmo Lekagne Joseph B, Nguefack J, Irénée S. Plant extracts induced resistance to Fusarium oxysporum f. sp. lycopersici in Tomato. International Journal of Plant and Soil Science. 2023;**35**:444-458. DOI:10.9734/ijpss/2023/v35i173229

[53] Sinno M, Ranesi M, Gioia L, D'errico G, Woo SL. Endophytic fungi of tomato and their potential applications for crop improvement. Agriculture. 2020;**10**:1-22. DOI: 10.3390/agriculture10120587

[54] Thambugala KM, Daranagama DA, Phillips AJL. Fungi vs. fungi in biocontrol: An overview of fungal antagonists applied against fungal plant pathogens. Frontiers in Cellular and Infection Microbiology. 2020;**10**:1-19. DOI: 10.3389/fcimb.2020.604923

[55] Samaras K, Mourtiadou S, Arampatzis T, Kakagianni M, Feka M, Wäckers F, et al. Plant-mediated effects of beneficial microbes and a plant Strengthener against spider mites in Tomato. Plants. 2023;**12**:1-13. DOI: 10.3390/plants12040938

[56] Daboy Dakole C, Nguefack J, Dongmo LJB, Galani YJH, Aza UR, Amvam ZPH. Antifungal potential of essential oils, aqueous and ethanol extracts of thirteen plants against Fusarium oxysporum f. sp Lycopersici and Phytophtora infestans (Mont.) de Bary as major tomato pathogens in Cameroon. International Journal of Current Science. 2016;**19**:128-145

[57] Ramyabharathi SA, Meena BL, Raguchander T. Induction of chitinase and β-1,3-glucanase PR proteins in tomato through liquid formulated Bacillus subtilis EPCO 16 against Fusarium wilt. The Journal Today's of Biology Scientific Research and Reviews. 2012;**1**:50-60

[58] Ascencio-Alvarez A, López-Benitez A, Borrego-Escalante F, Rodríguez-Herrera SA, Flores-Olivas A, Jiménez-Díaz F, et al. Marchitez vascular del tomate: I. Presencia de razas de Fusarium oxysporum f. sp. lycopersici (Sacc.) Snyder y Hansen en Culiacán, Sinaloa, México. Rev. Mex. Fitopatol. 2008;**26**:114-120

[59] Singh VK, Singh HB, Upadhyay RS. Role of fusaric acid in the development of 'Fusarium wilt' symptoms in

tomato: Physiological, biochemical and proteomic perspectives. Plant Physiology and Biochemistry. 2017;**118**:320-332. DOI: 10.1016/j.plaphy.2017.06.028

[60] Laurence MH, Summerell BA, Burgess LW, Liew ECY. Genealogical concordance phylogenetic species recognition in the Fusarium oxysporum species complex. Fungal Biology. 2014;**118**:374-384. DOI: 10.1016/j.funbio.2014.02.002

[61] Edel-Hermann V, Lecomte C. Current status of fusarium oxysporum formae speciales and races. Phytopathology. 2019;**109**:512-530. DOI: 10.1094/PHYTO-08-18-0320-RVW

[62] Vawdrey LL, Peterson RA. Fusarium solani, the cause of foot rot of tomatoes in Central Queensland. Australasian Plant Pathology. 1988;**17**:24-25. DOI: 10.1071/APP9880024

[63] Massee G. The "sleepy disease" of tomatoes. The Gardeners' Chronicle. 1895;**3**:705-707

[64] Swett CL, Del Castillo Múnera J, Hellman E, Helpio E, Gastelum M, Lopez Raymundo E, et al. Monitoring for a new I3 resistance gene-breaking race of F. Oxysporum f. sp. lycopersici (Fusarium wilt) in California processing tomatoes following recent widespread adoption of resistant (F3) cultivars: Challenges with race 3 and 4 differentiation metho. Frontiers in Plant Science. 2023;**14**:1-15. DOI: 10.3389/fpls.2023.1088044

[65] Cai G, Gale LR, Schneider RW, Kistler HC, Davis RM, Elias KS, et al. Origin of race 3 of Fusarium oxysporum f. sp. lycopersici at a single site in California. Phytopathology. 2003;**93**:1014-1022. DOI: 10.1094/PHYTO.2003.93.8.1014

[66] Rosa RCT, Gurgel LMS, Karasawa M, Andrade DETG, Assis TC. Detecção de Fusarium oxysporum f. sp. lycopersici em sementes de tomateiro pelo método de papel de filtro com adição de restritores. Pesqui. Agropecuária Pernambucana. 2022;**27**:e2692272022. DOI: 10.12661/pap.2022.006

[67] Hartman JR, Fletcher JT. Fusarium crown and root rot of tomatoes in the UK. Plant Pathology. 1991;**40**:85-92. DOI: 10.1111/j.1365-3059.1991.tb02296.x

[68] Leslie JF, Summerell BA. The Fusarium Laboratory Manual. 1st ed. Iowa, USA: Blackwell Publishing; 2006. DOI: 10.1002/9780470278376. ISBN 978-0-8138-1919-8

[69] Hibar K, Daami-Remadi M, Ayed F, El Mahjoub M. Fusarium crown and root rot of Tomato and its chemical control. International Journal of Agricultural Research. 2007;**2**:687-695. DOI: 10.3923/ijar.2007.687.695

[70] Ozbay N, Newman SE. Fusarium crown and root rot of tomato and control methods. Plant Pathology Journal. 2004;**3**:9-18. DOI: 10.3923/ppj.2004.9.18

[71] Jarvis WR. Taxonomic status of Fusarium oxysporum causing foot and root rot of Tomato. Phytopathology. 1978;**68**:1679

[72] Vega-Gutiérrez TA, López-Orona CA, López-Urquídez GA, Velarde-Félix S, Amarillas-Bueno LA, Martínez-Campos AR, et al. Foot rot and wilt in Tomato caused by Fusarium falciforme (FSSC 3 + 4) in Mexico. Plant Disease. 2018;**103**:157. DOI: 10.1094/PDIS-06-18-1001-PDN

[73] Summerbell RC, Schroers HJ. Analysis of phylogenetic relationship of Cylindrocarpon lichenicola and Acremonium falciforme to the Fusarium solani species complex and a review of similarities in the spectrum of opportunistic infections caused by these

fungi. Journal of Clinical Microbiology. 2002;**40**:2866-2875. DOI: 10.1128/JCM.40.8.2866-2875.2002

[74] Romberg MK, Davis RM. Host range and phylogeny of Fusarium solani f. sp. eumartii from potato and tomato in California. Plant Disease. 2007;**91**:585-592. DOI: 10.1094/PDIS-91-5-0585

[75] Silva SGA, Costa MM, Cardoso AMS, Nascimento LV, Barroso KA, Nunes GHS, et al. Fusarium falciforme and Fusarium suttonianum cause root rot of melon in Brazil. Plant Pathology. 2023;**72**:721-730. DOI: /10.1111/ppa.13701

[76] Sousa ES, Melo MP, Mota JM, Sousa EMJ, Beserra JEA Jr, Matos KS. First report of Fusarium falciforme (FSSC 3+ 4) causing root rot in lima bean (Phaseolus lunatus L.) in Brazil. Plant Disease. 2017;**101**:1954. DOI: 10.1094/PDIS-05-17-0657-PDN

[77] Tirado-Ramírez MA, López-OronaCA,deVelázquez-AlcarazTJ, Díaz-Valdés T, Velarde-Félix S, Martínez-Campos AR, et al. First report of onion basal rot caused by Fusarium falciforme in Mexico. Plant Disease. 2018;**102**:2646. DOI: /10.1094/PDIS-05-18-0757-PDN

[78] Vega-Gutiérrez TA, Tirado-Ramírez MA, López-Urquídez GA, Angulo-Castro A, Martínez-Gallardo JA, López-Orona CA. Fusarium falciforme (FSSC 3+ 4) causing root and stem rot in papaya (Carica papaya) in Mexico. Plant Disease. 2019;**103**:2681. DOI: /10.1094/PDIS-05-19-0917-PDN

[79] Balasubramaniam J, Goh KS, Sani SF, Alam MW, Ismail NA, Gleason ML, et al. Fusarium falciforme and F. Oxysporum causing postharvest fruit rot of watermelon (Citrullus lanatus) in Malaysia: A first report. Crop Protection. 2023;**163**:106115. DOI: /10.1016/j.cropro.2022.106115

[80] O'donnell K. Molecular genetic approaches for measuring microbial diversity: Examples from filamentous fungi. In: Proceedings of the Diversity and Use of Agricultural Microorganisms: The 5th International Workshop on Genetic Resources. Tsukuba, Ibaraki, Japan: National Institute of Agrobiological Resources; 1998. pp. 7-24

[81] Edupuganti S, Rouphael N, Mehta A, Eaton M, Heller JG, Bressler A, et al. Fusarium falciforme vertebral abscess and osteomyelitis: Case report and molecular classification. Journal of Clinical Microbiology. 2011;**49**:2350-2353. DOI: 10.1128/JCM.02547-10

[82] López-López AM, Allende-Molar R, Correia KC, Tovar-Pedraza JM, Márquez-Zequera I, García-Estrada RS. First report of stem canker of Tomato caused by Fusarium striatum in Mexico. Plant Disease. 2020;**105**:497. DOI: 10.1094/PDIS-06-20-1187-PDN

[83] Guimarães SSC, Matteoli FP, Pedrosa-Silva F, Matos KS, Pfenning LH. Structural organization of the mating-type locus in clade 3 of Fusarium solani species complex and development of PCR protocol for identification of MAT idiomorphs. Current Research in Environmental & Applied Mycology. 2021;**11**:159-170. DOI: 10.5943/cream/11/1/12

[84] Zhou X, Luo C, Li K, Zhu D, Jiang L, Wu L, et al. First report of fusarium striatum causing root rot disease of panax notoginseng in Yunnan, China. Phyton-International Journal of Experimental Botany. 2022;**91**:13-20. DOI: 10.32604/phyton.2022.016972

[85] Fajola AO. The post-harvest fruit rots of Tomato (Lycopersicum esculentum) in Nigeria. Food/Nahrung. 1979;**23**:105-109. DOI: 10.1002/food.19790230202

[86] Zhao Z, Yang X, Zhao X, Chen L, Bai B, Zhou C, et al. Method

development and validation for the analysis of emerging and traditional Fusarium mycotoxins in pepper, potato, Tomato, and cucumber by UPLC-MS/MS. Food Analytical Methods. 2018;**11**:1780-1788. DOI: 10.1007/s12161-018-1180-7

[87] Muhammad S, Shehu K, Amusa NA. Survey of the market diseases and aflatoxin contamination of tomato (MILL) fruits in Sokoto, northwestern Nigeria. Nutrition & Food Science. 2004;**34**:72-76. DOI: 10.1108/00346650410529032

[88] Meca G, Font G, Ruiz MJ. Comparative cytotoxicity study of enniatins a, A1, A2, B, B1, B4 and J3 on Caco-2 cells, hep-G2 and HT-29. Food and Chemical Toxicology. 2011;**49**:2464-2469. DOI: /10.1016/j.fct.2011.05.020

[89] Rabie CJ, Marasas WF, Thiel PG, Lübben A, Vleggaar R. Moniliformin production and toxicity of different Fusarium species from southern Africa. Applied and Environmental Microbiology. 1982;**43**:517-521. DOI: 10.1128/aem.43.3.517-521.1982

[90] Marín S, Ramos AJ, Cano-Sancho G, Sanchis V. Reduction of mycotoxins and toxigenic fungi in the mediterranean basin maize chain. Phytopathologia Mediterranea. 2012;**51**:93-118. DOI: 10.14601/Phytopathol_Mediterr-9411

[91] Awuchi CG, Ondari EN, Nwozo S, Odongo GA, Eseoghene IJ, Twinomuhwezi H, et al. Mycotoxins' toxicological mechanisms involving humans, livestock and their associated health concerns: A review. Toxins (Basel). 2022;**14**:1-33. DOI: 10.3390/toxins14030167

[92] Ismail AM, Elshewy ES, El-Ganainy SM, Magistà D, Hamouda AF, Alhudaib KA, et al. Mycotoxins from Tomato pathogenic Alternaria alternata and their combined cytotoxic effects on human cell lines and male albino rats. Journal of Fungi. 2023;**9**:1-20. DOI: 10.3390/jof9030282

[93] Ruiz MJ, Franzova P, Juan-García A, Font G. Toxicological interactions between the mycotoxins beauvericin, deoxynivalenol and T-2 toxin in CHO-K1 cells in vitro. Toxicon. 2011;**58**:315-326. DOI: 10.1016/j.toxicon.2011.07.015

[94] Prosperini A, Font G, Ruiz MJ. Interaction effects of Fusarium enniatins (a, A1, B and B1) combinations on in vitro cytotoxicity of Caco-2 cells. Toxicology in vitro. 2014;**28**:88-94. DOI: 10.1016/j.tiv.2013.06.021

[95] Fumagalli F, Ottoboni M, Pinotti L, Cheli F. Integrated mycotoxin management system in the feed supply chain: Innovative approaches. Toxins (Basel). 2021;**13**:1-35. DOI: 10.3390/toxins13080572

[96] Assefa T, Geremew T. Major mycotoxins occurrence, prevention and control approaches. Biotechnology and Molecular Biology Reviews. 2018;**12**:1-11. DOI: 10.5897/BMBR2018.0274

[97] Akram W, Anjum T, Ahmad A. Basal susceptibility of tomato varieties against different isolates of fusarium oxysporum f. sp. lycopersici. International Journal of Agriculture and Biology. 2014;**16**:171-176

[98] Yan H, Nelson B. Effects of soil type, temperature, and moisture on development of Fusarium root rot of soybean by Fusarium solani (FSSC 11) and Fusarium tricinctum. Plant Disease. 2022;**106**:2974-2983. DOI: 10.1094/PDIS-12-21-2738-RE

[99] Blanco R, Aveling TAS. Seed-borne Fusarium pathogens in

agricultural crops. Acta Horticulturae. 2018;**1204**:161-169. DOI: 10.17660/ActaHortic.2018.1204.21

[100] Chitwood-Brown J, Vallad GE, Lee TG, Hutton SF. Breeding for resistance to fusarium wilt of tomato: A review. Genes (Basel). 2021;**12**:1673. DOI: 10.3390/genes12111673

[101] Bawa I. Management strategies of fusarium wilt disease of tomato incited by *Fusarium oxysporum* f. sp. lycopersici (Sacc.): A review. International Journal of Advanced Academic Research | Science Technology and Engineering. 2016;**2**:2488-9849

[102] Pritesh P, Subramanian RB. PCR based method for testing Fusarium wilt resistance of Tomato. African Journal of Basic & Applied Sciences. 2011;**3**:219-222

[103] Song F, Goodman RM. Molecular biology of disease resistance in rice. Physiological and Molecular Plant Pathology. 2001;**59**:1-11. DOI: 10.1006/pmpp.2001.0353

[104] Prihatna C, Barbetti MJ, Barker SJ. A novel tomato Fusarium wilt tolerance gene. Frontiers in Microbiology. 2018;**9**:1-11. DOI: 10.3389/fmicb.2018.01226

[105] Patel S, Saraf M. Biocontrol efficacy of Trichoderma asperellum MSST against tomato wilting by Fusarium oxysporum f. sp. lycopersici. Archives of Phytopathology and Plant Protection. 2017;**50**:228-238. DOI: 10.1080/03235408.2017.1287236

[106] Raza W, Ling N, Zhang R, Huang Q, Xu Y, Shen Q. Success evaluation of the biological control of Fusarium wilts of cucumber, banana, and tomato since 2000 and future research strategies. Critical Reviews in Biotechnology. 2017;**37**:202-212. DOI: 10.3109/07388551.2015.1130683

[107] Wang L, Liu J, Nie Y, Wang D, Wang H. Effects of vermicompost on tomato Fusarium wilt and soil microbial community structure. Acta Agriculturae Scandinavica. Section B. Soil and Plant Science. 2021;**71**:835-851. DOI: 10.1080/09064710.2021.1952301

[108] Someya N, Tsuchiya K, Yoshida T, Noguchi MT, Sawada H. Combined use of the biocontrol bacterium *Pseudomonas fluorescens* strain LRB3W1 with reduced fungicide application for the control of Tomato Fusarium wilt. Biocontrol Science. 2006;**11**:75-80. DOI: 10.4265/bio.11.75

[109] Ajilogba CF, Babalola OO, Ahmad F. Antagonistic effects of bacillus species in biocontrol of tomato Fusarium wilt. Studies on Ethno-Medicine. 2013;7:205-216. DOI: 10.1080/09735070.2013.11886462

[110] Baysal Ö, Çalışkan M, Yeşilova Ö. An inhibitory effect of a new Bacillus subtilis strain (EU07) against Fusarium oxysporum f. sp. radicis-lycopersici. Physiological and Molecular Plant Pathology. 2008;**73**:25-32. DOI: 10.1016/j.pmpp.2008.11.002

[111] Myresiotis CK, Karaoglanidis GS, Vryzas Z, Papadopoulou-Mourkidou E. Evaluation of plant-growth-promoting rhizobacteria, acibenzolar-S-methyl and hymexazol for integrated control of Fusarium crown and root rot on tomato. Pest Management Science. 2012;**68**:404-411. DOI: 10.1002/ps.2277

[112] Khalil MMR, Fierro-Coronado RA, Peñuelas-Rubio O, Villa-Lerma AG, Plascencia-Jatomea R, Félix-Gastélum R, et al. Rhizospheric bacteria as potential biocontrol agents against Fusarium wilt and crown and root rot diseases in tomato. Saudi Journal of Biological Sciences. 2021;**28**:7460-7471. DOI: 10.1016/j.sjbs.2021.08.043

[113] Sivan A. Biological control of Fusarium crown rot of Tomato by Trichoderma harzianum under field conditions. Plant Disease. 1987;**71**:587

[114] Datnoff LE, Nemec S, Pernezny K. Biological control of Fusarium crown and root rot of Tomato in Florida using Trichoderma harzianum and glomus intraradices. Biological Control. 1995;**5**:427-431. DOI: 10.1006/bcon.1995.1051

[115] Malandrakis A, Daskalaki ER, Skiada V, Papadopoulou KK, Kavroulakis N. A Fusarium solani endophyte vs fungicides: Compatibility in a Fusarium oxysporum f. sp. radicis-lycopersici – Tomato pathosystem. Fungal Biology. 2018;**122**:1215-1221. DOI: 10.1016/j.funbio.2018.10.003

[116] Muslim A, Horinouchi H, Hyakumachi M. Control of Fusarium crown and root rot of Tomato with Hypovirulent Binucleate Rhizoctonia in soil and rock wool systems. Plant Disease. 2003;**87**:739-747. DOI: 10.1094/PDIS.2003.87.6.739

[117] Khan S, Khan R. Eco-friendly management strategy against fusarium wilt of tomato caused by Fusarium oxysporum f. sp lycopersici (SACC.). Pakistan Journal of Weed Science Research. 2022;**28**:479-486. DOI: 10.28941/pjwsr.v28i4.1099

[118] Mostafa YS, Alamri SA, Alrumman SA, Hashem M, Taher MA, Baka ZA. In vitro and In vivo biocontrol of Tomato Fusarium wilt by extracts from Brown, red, and green macroalgae. Agriculture. 2022;**12**:1-18. DOI: 10.3390/agriculture12030345

[119] El-Aswad AF, Aly MI, Alsahaty SA, Basyony ABA. Efficacy evaluation of some fumigants against Fusarium oxysporum and enhancement of tomato growth as elicitor-induced defense responses. Scientific Reports. 2023;**13**:1-12. DOI: 10.1038/s41598-023-29033-w

[120] Li X, Wang Q, Li H, Wang X, Zhang R, Yang X, et al. Revealing the mechanisms for linalool antifungal activity against Fusarium oxysporum and its efficient control of Fusarium wilt in Tomato plants. International Journal of Molecular Sciences. 2023;**24**:458. DOI: 10.3390/ijms24010458

[121] Borrego-Benjumea A, Basallote-Ureba MJ, Abbasi PA, Lazarovits G, Melero-Vara JM. Effects of incubation temperature on the organic amendment-mediated control of Fusarium wilt of tomato. The Annals of Applied Biology. 2014;**164**:453-463. DOI: 10.1111/aab.12116

[122] Zhao F, Zhang Y, Dong W, Zhang Y, Zhang G, Sun Z, et al. Vermicompost can suppress Fusarium oxysporum f. sp. lycopersici via generation of beneficial bacteria in a long-term tomato monoculture soil. Plant and Soil. 2019;**440**:491-505. DOI: 10.1007/s11104-019-04104-y

[123] Szczech MM. Suppressiveness of vermicompost against fusarium wilt of tomato. Journal of Phytopathology. 1999;**147**:155-161. DOI: 10.1046/j.1439-0434.1999.147003155.x

Chapter 6

Effects of *Fusarium* Diseases on Forest Nursery and Its Controlling Mechanisms

Adugnaw Mekonene Admas and Leykun Kebede

Abstract

Fusarium spp. cause severe harm to forest plants. These fungi can induce latent infections that lead to planted seedlings rooting, as well as pre- and post-emergence damping. However, a phylogenetic study suggests that the most virulent strains of *Fusarium oxysporum Schlechtend*, which has been identified as the primary cause of root and root rot in nurseries, are more closely aligned with the recently identified *Fusarium genus, Skovgaard, O'Donnell, and Nirenberg*. Before planting, soil fumigation was the primary method of treating Fusarium illnesses in nurseries with bare roots. Alternative therapies are being investigated as rules impede the supply of the most effective fumigants. This entails improving sanitation, preserving a healthy microbial population that inhibits pathogens infecting trees through their roots, and refraining against actions that increase the risk of disease in trees, such as overfertilization and inadequate soil drainage. Although *Fusarium circinatum, Nirenberg*, and *O'Donnell* can be problematic in nurseries, they can harm ancient trees in native forests, plantations, seed farms, and landscape plantings. It is not advisable to transfer seeds or seedlings from contaminated to uninfected areas because they can spread the virus. To stop *F. circinatum* from spreading to nations where it has not yet been discovered, quarantine measures must be upheld. *F. circinatum* infections are linked to harm from weather-related events, insect activity, pruning, and seed harvesting, among other forest management practices. Pruning during the cold, dry season, when conditions are less conducive to infection, can help minimize the risk of illness in managed plantations and control insects that have the potential to be vectors and pests. Ecologically friendly biological strategies, such as using endophytic fungi and bacteria that are antagonistic to *F. circinatum*, plant essential oils, chitosan, or phosphite, have also been researched as ways to lessen the impact. Additionally, to reduce the number of contaminated seeds introduced into nurseries in disease-free areas, heat treatment is an easy and affordable way to eradicate the pathogen from contaminated seeds. Therefore, to address the problems of *Fusarium spp*,effects on nurseries, natural forests, and plantations using integrated approaches is required for sustainable managements of the forests.

Keywords: fungi, forest trees, *Fusarium*, treatment, diseases

1. Introduction

Most of the world's surface area is covered by forests, which are essential ecosystems that play a major role in the global carbon cycle, supply valuable economic products, and provide ecosystem services. Approximately 112.95 million hectares of Europe's forest area are covered by conifer trees, primarily Pinus species [1].

Pinus is a highly significant tree in terms of both ecology and economics. Furthermore, they are a major supplier of wood, pulp and paper, edible seeds, charcoal, resin, and many other biological activities in forest ecosystems [2]. However, free-market policies and commercial globalization have brought forest infections, particularly dangerous alien species, as invasives into any nation. This causes serious environmental and social impacts and economic losses globally [3–5].

In this regard, plant protection regulations often fail to prevent biological invasions [6–8]; and as a result, the number of invasive and damaging forest pathogens has increased dramatically recently [9].

To cope with this global challenge, new pest and disease management approaches are urgently needed to preserve the diverse services of forests under the increasing risks of globalization and climate change [10]. Fusarium species are among the ecologically harmful fungi that can infiltrate the soil and give older trees or young seedlings symptoms of neck and root rot. On the other hand, little is known about the diseases that afflict trees or seedlings in natural and semi-natural forests.

Fusarium species were found in the roots of oak (*Quercus robur L.*), white oak (*Q. petraea (Matt.) Liebl.*), and ash (*Fraxinus Excelsior L.*) in the former Czechoslovakia [11], Austria [12], and Poland [13]. Numerous ecological niches, including those for pathogens, endophytes, and saprotrophs, are occupied by the genus Fusarium in various climatic zones, particularly in tropical and subtropical areas [14]. This species is deemed dangerous primarily because it seriously harms forestry, which restricts seedling output globally [15], as seen with *Fusarium circinatum Nirenberg* in numerous coniferous *O'pine species.* Donnell can result in significant drops in the productivity and quality of fruits, vegetables, and other edible plant materials, particularly in South Africa [16–19]. Fungal pathogen-caused plant illnesses are becoming more and more noticeable to consumers today [16–18, 20].

Fusarium species cause inhibition of germination, root rot, seedling wilt, stem rot, and seed rot in orchards. Many species of Fusarium, *including F. oxysporum sensu lato., F. parish K. Skovg., F. equiseti (Corda) Sacc., F. chlamydosporum Wolverine., F. circinatum Nirenberg ja O'Donnell, F. proliferatum (Matsush.) Nirenberg., F. acuminatum Ellis kaj Everh. Vaca Reinking., F. moniliforme J. Shield, and F. tricinctum (Corda) Sacc.* Young seedlings in nurseries and germination-prone seeds are susceptible to infection by Necosmospora solani (Mart.) L. Lombard and Crous (formerly known as *Fusarium solani (Mart.) Sacc.).* The *oxysporum sensu lato* species complex poses the greatest threat among *Fusarium* members [16–18, 21–24].

It is unknown if Fusarium species are associated with the spontaneous regrowth of forest trees, despite the fact that members of this type of phytopathogenic fungus have been identified in forest litter and soil. For example, *F. avenaceum and F. lateritium* were discovered in oak and sycamore (*Acer pseudoplatanus* L.) in Great Britain [25], whereas F. proliferatum was discovered in the soil of pine forests (*Pinus sylvestris* L.) in Scotland [26] and Poland [27]. No representatives of the genus *Fusarium* are more frequently discovered in soil samples taken from open fields than in samples taken from regenerating and mature forests. Outside Europe, *Fusarium spp.* was found in the roots of naturally regenerating Douglas fir (*Pseudotsuga menziesii* (Mirb.) Franco) seedlings in Canada [28, 29] and in Japanese hardwood seedlings with declining symptoms [30]. Pathogenic *Fusarium spp.* were reported to be associated with tree roots and rhizomes in western Iran [31], whereas *F. avenaceum and Freziera longipes Wollenw and Reinking* were found as inhibitory pathogens in naturally regenerating *Eucalyptus obliqua L'Hértier* and *E. radiata in Sieber ex DC.* in Australia [32]. Recently, *F. solani, F. oxysporum, F. verticillioides (Sacc.) Nirenberg, F. equiseti, F. fujikuroi Nirenberg, F. pseudocircinatum O'Donnell Ning Nirenberg kaj F. subglutinans (Wollenw. and Reinking) P.E. Nelson, Toussoun, and Marasas* having the potential to cause leaf branch darkening were confirmed in pathogenicity tests [33].

Members of the genus *Fusarium* are commonly found in soil and forests in different parts of the world. For instance, populations of different *Fusarium species* have been reported in forests of Canada [34], Australia [35], Sri Lanka [36], India [37], Malaysia [38], Argentina [39], and Indonesia [40]. Fungal pathogens living in the litter layer can reduce the number of budding beech trees. According to a recent study, *Neonectria and Ilyonectria species*, which inhabit beech beds, are pathogenic and can harm *F. sylvatica's* normal regeneration [41]. Numerous members of the genus *Fusarium* are more common, including unexplored species linked to the early stages of forest trees, according to a 2011 investigation of beech soil fungi in Central Europe [42]. Hence, this book chapter focuses more on Fusarium sp. effects on forest nurseries and its controlling mechanisms.

2. Causative agents of Fusarium fungi and others

Phytophthora, Pythium, Fusarium, and *Rhizoctonia* are the most frequent causes of plant growth inhibition [43]. However, **Table 1** below includes fungi from a number of other genera, such as *Colletotrichum, Alternaria,* and *Cylindrocladium*. Commonly, *Rhizoctonia spp*. have been found to be significant contributors to damping-off in ornamental orchards [60, 61] and to be a source of disease in tree seedlings. It is not isolated more often in the US; it may be that its presence is masked by faster-growing fungi such as *Fusarium spp* [62]. Recent literature [63] lists the most common damping-off pathogens.

Pathogen	***Host sp***	**County**	**Source**
Fusarium spp.	*Pinus nigra*	Spain	[44]
etotrichum acutatum, Fusarium oxysporum	*Cornus florida*	USA: Georgi	[45]
Alternaria tenuis, Fusarium spp., Pythium spp., Rhizoctonia solani	*Eucalyptus spp.*	China	[46]
Rhizoctonia solani	*Caragana arborescens*	Canada	[47]
Cylindrocladium scoparium, Rhizoctonia solani	*Eucalyptus spp.*	Brazil	[48]
Rhizoctonia spp, Pythium spp	*Picea smithiana*	India	[49]
Rhizoctonia spp.	*Pinus palustris*	USA: Florida	[50]
Fusarium spp.	*Pinus sylvestris*	Finland	[51]
Fusarium spp., Pythium spp., Thanatephorus spp	*Eucalyptus spp., Pinus caribaea, Acacia spp.*	Zimbabwe	[52]

Pathogen	*Host sp*	County	Source
Phoma herbarum, Phomopsis occulta	*Larix decidua*	France	[53]
Fusarium spp., Phytophthora spp., Rhizopus spp.	*Santalum album*	India	[54]
Colletotrichum dematium	*Fagus crenata*	Japan	[55]
Fusarium spp.	*Pseudotsuga*	USA: Idaho	[56]
Phytophthora spp.	*Fagus sylvatica*	Poland	[57]
Cylindrocarpon destructans	*Pinus sylvestris*	Sweden	[58]
Cylindrocladium scoparium	*Pinus resinosa*	Canada	[41]
Fusarium spp., Pythium spp., Rhizoctonia spp	*Acacia mangium*	Phillipines	[59]
Fusarium spp., Alternaria spp., Pythium spp.	*Pinus sylvestris, Larix silbirica*	Russia	[48]

Table 1.
Damping-off is a cosmopolitan disease affecting plants from around the world.

3. Fusarium fungi disease control

Fusaria pathogens can be carried out by the soil and are extremely difficult to eliminate once they have taken over a field. Currently, fumigants such as chloropicrin, hot water, sunlight, or cultivars that are resistant to the pathogen are commonly employed to disinfect the soil and fight diseases caused by soil-borne pathogens.

Unfortunately, the results of these treatments are not as strong as anticipated, will harm the environment, and necessitate reducing their use.

Biological control refers to the use of microorganisms in the management of plant diseases. To combat tobacco *Rhizoctonia*, for instance, *Trichoderma lignorum* was registered as a fungicide in 1954 under Japan's Agricultural Chemicals Regulation Act. This was the world's first instance of biofungicides being documented. Many investigations were carried out to accomplish biocontrol of *Fusarium* illnesses throughout the last 3 to 4 decades, including the rice-causing Bakanae disease, which is brought on by soil-borne *F. oxysporum* and *F. fujikuroi*. As a biofungicide against *F. oxysporum sp* on sweet potatoes, a nonpathogenic strain of *F. oxysporum* was registered in 2002. However, *Talaromyces flavus* was registered in 2007 after *Trichoderma atroviride* was registered in 2003 as a biofungicide to control Bakanae rice through seed or nursery treatment. They are both available in the market currently. Furthermore, for plant health certification and effective disease management, soil cleaning, the use of disease-resistant cultivars, the application of fungicides, and the early detection of plant pathogens in the field are crucial. Nevertheless, there was no practical procedure for Fusarium spp. except for the isolation and inoculation test with test plants.

Seeds and seedlings should not be transferred from contaminated to uninfected areas because they can spread the disease. To stop *F. circinatum* from spreading to nations where it has not yet been discovered, quarantine measures must be upheld. *F. circinatum* infections are linked to harm from weather-related events, insect activity, pruning, and seed harvesting, among other forest management practices. Pruning should only take place during the cold, dry season when conditions are less conducive to infection. Additionally, insects that can serve as pests and vectors should be carefully controlled to lower the risk of diseases in managed plantations. Furthermore, heat treatment is an easy and affordable way to get rid of the pathogen from contaminated seeds to reduce the entry of the infected seeds into nurseries of disease-free areas.

DOI: http://dx.doi.org/10.5772/intechopen.1004156

4. Type of pest and method of propagation

The most common culprits are several genera of fungi (*Fusarium*, *Rhizoctonia*) and *Oomycetes* (*Pythium, Phytophthora*), but other fungi can also be involved **(Table 1).** If general controls are not effective, the presence of the pathogen must be confirmed by culturing it in an artificial medium [64]. However, the way of propagation is very different for each pest. Dispersal in contaminated soil or growing medium is typical for all species. *Fusarium spp.* spores are spread through contaminated seeds, soil, or growing medium and used containers. The role of seed dispersal is readily apparent in cotyledon blight. Although spores are airborne, they are primarily responsible for secondary spread. Thick-walled chlamydos spores help the fungus overwinter in plant litter, and sclerotia also can form [65]. *Rhizoctonia sp.* can be spread by seeds or airborne spores, but transmission is most common through contaminated soil, as the fungus overwinters in the soil as sclerotia [66]. Oocyst producers of *Pythium spp*. and *Phytophthora spp*. have unique zoospores that can swim in water and overwinter in soil or plant debris such as thick-walled oospores or chlamydospores. None of these pathogens produce airborne spores, although spores can be spread by water spray [67].

5. Critical control points for nursery damping

The best management method is to keep pathogens out of the nursery; however, this isnot always feasible, particularly in nurseries with bare roots where diseases can still survive in the soil. Because benches, containers, and other surfaces are easily sanitized, seedling preservation is made easier from the perspective of disease prevention [68], mostly diseases that inhibit seeds, such as Fusarium species and Rhizoctonia species. **Table 1** also lists additional, less aggressive fungi that can be problematic for certain species. To remove this possible source of inoculum, seeds are rinsed under running water or sterilized with a diluted (1:10) solution of hydrogen peroxide or chlorine before being sown [69]. Because all suppressive pathogens are frequent soil dwellers, they can readily get caught in-between crops or develop into dormant spores that might linger for months or even years in plant debris. Therefore, sterilizing the soil before sowing would be a wise management practice for bare-root nurseries.

One of the main ways that Rhizoctonia spreads in bare-root at nurseries is through soil splashing, which can be reduced by plant cushions [70]. The majority of media used in container gardens, such as perlite and vermiculite, are practically sterile, and pathogen suppression is hindered by the low pH of sphagnum peat [71]. The most typical way that *Pythium* and *Phytophthora* spread is through water because of their motile zoospores. Reducing surface water contact and polluted runoff by storing containers on raised benches is advised to utilize a raised bed in bare-root nurseries to prevent water from standing around the seedlings and to select coarser-textured, well-drained soil for seed sowing [72].

6. The role of the environment

Because most suppressive organisms are opportunistic pathogens, using the right culture techniques can lessen or even eliminate illness. For more than a century, water logging can be avoided by simply maintaining low pH levels in the soil or growing medium. Another way to avoid wetness is to keep the soil or growing medium "moist but not wet" [73].

6.1 Impact of climate change on *Fusarium circinatum* and other plant pathogens

A virulent infection, a susceptible host plant, and ideal environmental circumstances work together to generate plant disease. Damage caused by diseases has a high correlation with environmental changes [74]. The intensity of the illness and the coevolution of pathogens and plants can be impacted by changes in temperature, precipitation, drought, and other factors [75, 76]. In addition to having a direct impact on pathogen biology, environmental variables have a significant impact on plant germination and reproduction. For foliar pathogens, these parameters include temperature, relative humidity, and leaf wetness [77, 78]. Besides these, temperature and relative humidity can affect pathogen overwintering in the absence of a host.

Because pathogens have evolved to withstand a wide range of temperatures and have evolved at high latitudes, it is anticipated that climate change will boost their fitness and raise the danger of disease outbreaks [79]. Reduced rainfall can lessen grape downing, but higher temperatures offset shorter leaf wetness durations because infections begin earlier in the growth season, giving epidemics more time to spread [80].

Pathogens such as *Fusarium circinatum* (pine canker), which is more common in Europe, are favored by a drop in the average minimum temperature accompanied by more frost [81].

Conversely, variables such as increased CO_2 concentrations, elevated temperatures, or dry spells impact host physiology, which modifies the way biotrophic infections colonize host tissues [82]. According to reports from central Italy, the invasive exotic species *Heterobasidion* irregular—which is more suited to spread in the Mediterranean climate than the local species *Heterobasidion annosum*—is directly impacted by dry circumstances [83].

Numerous diseases differ depending on the temperature and humidity. Because they can spread swiftly by wind and have shorter generations than plants, most diseases have an advantage over plants that allows them to move and adapt to changing climate conditions [84].

Changes in the flora of some fungus appear to be linked to climate change, which poses a threat to plant health and, in certain situations, may also increase the prevalence of pests and diseases. Infections with *Phytophthora infestans* have increased in Finland as a result of climate change [85]. In Canada and the United States, high temperatures have been linked to significant beetle outbreaks [20, 86, 87]. Models indicate that the number of canola diseases in Scotland will decline due to *Pyrenopeziza brassicas* and *Leptosphaeria maculans* [88]. Nonetheless, in the long run (2071–2100), increased average temperatures in northern Germany are anticipated to benefit canola pathogens such *Alternaria brassicae, Sclerotinia sclerotiorum, and Verticillium longisporum* [89]. Although there is currently little knowledge on how climate change may affect plant disease biological control, high temperatures may, in certain situations, make the introduction of biological control easier [90, 91]. Crops, fruits, and other edible plant materials can suffer significant yield and quality losses due to plant diseases. In contemporary times, fungal pathogen-caused plant diseases have garnered the interest of consumers as a significant global health concern [92].

7. *Fusarium* wilt, root rot, and disease

It has been documented that suppressing pathogenic fungi such as *Fusarium oxysporum*, *R. Solanum*, *and Pythium aphanidermatum* can lead to damping-off illnesses, root rot, and Fusarium wilt in a variety of crops, which can seriously reduce growth and yield [93].

Endophytic bacteria that colonize plant interiors have even been demonstrated in recent research to enhance plant growth and health [94]. For this reason, they seem to be great candidates as biocontrol agents against fungal phytopathogens. Three endophytic bacterial strains were assessed for their antifungal qualities: *B.B. Simbolis H18 (Bacillales: Bacillaphy), Pseudomonas aeruginosa H40 (Pseudomonadales: Pseudomonadaceae), and Stenotrophomonas maltophila H8 (Xanthomonadales: Xanthomonadaceae).* By treating plants with endophytic bacteria, they were able to lessen the severity of their diseases and boost seed production and seedling survival.

Furthermore, treated seedlings exhibit an apparent increase in the antioxidant enzymes peroxidase, PPO, and catalase (CAT). The capacity of bacterial endophytes to generate a broad range of antifungal chemicals, in addition to bioactive substances, which can cause systemic resistance in infected cotton seeds, has been linked to their antagonistic action against *R. solani* [95].

Maximum chitinase production was seen in *B. thuringiensis* GS1, an endophytic bacterium that was isolated from *Pteridium aquilinum*, a rough fern. It was also investigated if *B. thuringiensis* GS1 may make cucumber plants resistant to *R. solani*. *R. solani* wilt in cucumber plants may be prevented by *B. thuringiensis* GS1 induction of PRPs and defense-related enzymes [96], as well as the use of endophytic actinomycetes to biocontrol *R. solani*, which caused tomato wilt disease [97]. The endophytic P. fluorescent strain ENPF1 inhibited the growth of the stem blight pathogen *Corynespora casiicola (Berk and Curt)* Wei in vitro. P. fluorescents increased defense-related enzymes such as peroxidase, PPO, chitinase, and β-1,3-glucanase in *Phyllanthus amarus* after exposure to *Corynespora cassiicola* [98].

8. Biological control of emerging forest diseases

8.1 Biological control of nonnative pathogens

Severe disease outbreaks can result from novel pathogen-host interactions. In actuality, nonnative viruses may have detrimental effects on nearby nonnative tree species as well as native tree species. For instance, Fusarium circinatum on Pinus species (mostly *P. radiata*) in the Northern and Southern Hemispheres, *P. pini* on *Pinus radiata* in Chile, and Phytophthora ramorum on Japanese larch (*Larix kaempferi*) in the British Isles [98]. The native location of the virus is more likely to host coevolved hyperparasites.

It is true that a freshly distributed pathogen epidemic that intensifies can alter the afflicted ecology significantly. Here, polyphagous species coexisting in the same ecosystem might eventually switch to a different carbon source and modify their biology to take advantage of the newcomer [99]. However, attempts have been made to biologically control several of the most devastating nonnative pathogens using a bottom-up strategy; many potential biological control agents (BCAs), although promising in vitro, have almost completely failed in the field (e. g., *Fusarium circinatum*; failure in situ, and this may be climatic constraints, lack of an alternative host, development of resistance to BCA, and biotic disturbance by native organisms).

Reinforcement of top-down control forces through the recognition of native forest allies is particularly attractive in a foreign species pathogen × (native or alien) host. In relation to this, interspecific variation in host resistance is usually limited, and coevolution does not occur. The effects of symbionts on host defenses must be general enough to protect against pathogens that have not evolved, according to Halecker et al. [100] who identified an endophyte (*Hypoxylon rubiginosum*) living in European ash (*Fraxinus excelsior*) leaves that produces metabolites toxic to the nonsymbiotic pathogen *Hymenoscyphus fraxineus*. Thus, this endophyte may have the potential to evolve into an effective BCA. In the inoculation experiment, it was found that the presence of a local (*Heterobasidion annosum*) or a foreign root pathogen (*H. irregulare*) on *Pinus pinea* seedlings also affected the mycorrhizal density in the same way. However, the *ectomycorrhizal symbiont* (Tuber borchii) appeared to be able to discriminate between native and nonnative pathogens, possibly through host plant-mediated signaling [100].

9. Treatment of *Fusarium* diseases on conifers

Numerous *Fusarium species* are thought to be responsible for conifer diseases, some of which have detrimental effects on the environment and economy. Among these are fungi that result in poor growth in transplanted seedlings, stunting, and root rot in nurseries [101]. *In addition to harming seedlings, Fusarium circinatum Nirenberg and O'Donnell are pathogens that affect adult trees, resulting in pitch canker disease. Along with stem and root rot brought on by Fusarium oxysporum and other Fusarium communities, tonal canker also affects planted pines. Fusarium torreyae T. by Aoki, J.A. Smith, L. Mountain, Geiser, and O'Donnell, a recently reported pathogen of Torreya taxifolia, are among the Fusarium species that have been linked to plant illnesses but have not yet been well explored* [102].

Tonal canker, which is caused by *F. circinatum*, was originally discovered to be a disease in North Carolina, U.S.A. in 1945 [103]. The disease was given the moniker "pitch cancer" because the pathogen was found in malignant tumors that produced a lot of resin. Later, pitch canker made its way to the U.S. Southeast, where it still poses a threat to pine plantations, seeds, and seedlings.

10. Conclusions

The majority of forest diseases, particularly dangerous alien species such as *Fusarium*, are brought into nonnative areas through free market policies and worldwide commerce. This has detrimental effects on the environment, society, and the forest industries' bottom line. Therefore, laws pertaining to plant protection sometimes fall short of preventing biological invasions, which has led to a sharp rise in the quantity of invasive and harmful forest pathogens in recent years. To cope with this worldwide challenge and protect the various services provided by forests from the growing threats of globalization and climate change, new methods of managing pests and diseases are desperately needed.

In various climatic zones of tropical and subtropical regions, the fungus Fusarium occupies multiple ecological niches, including those of pathogens, endophytes, and saprotrophs. In particular, Fusarium circinatum Nirenbergi in many coniferous species of the O'pine genus is seen to be a problem primarily because it severely harms

DOI: http://dx.doi.org/10.5772/intechopen.1004156

globally by restricting the generation of seedlings. In addition, Fusarium circinatum Nirenbergi has the potential to significantly lower fruit, crop, and other edible plant material yields as well as quality.

Hence, to prevent this fungus from damaging nurseries, before sowing, the seeds are cleaned under running water or sterilized with a diluted (1:10) solution of chlorine or hydrogen peroxide that can eliminate this potential source of inoculums.

High latitudes are where pathogens evolve. Because of the pathogen's ability to withstand a broad range of temperatures, it is anticipated that climate change may boost their fitness and raise the possibility of disease outbreaks, especially those involving Fusarium species. The severity of the disease and the coevolution of pathogens and plants can be influenced by variations in temperature, precipitation, drought, and other environmental factors. These changes are closely linked to disease-induced damage. Several species of *Fusarium* cause diseases of conifers, some of which have serious economic and ecological impacts. These include fungi that cause stunting and root rot in nurseries and poor growth of transplanted seedlings.

Furthermore, reducing the chance of importing contaminated seeds into nurseries in disease-free areas can be achieved easily and affordably by using thermotherapy to eradicate the virus from infected seeds. However, *Fusarium spp*. will remain an unsolved problem for the foreseeable future due to the absence of practical field intervention strategies and the challenges of preventing its spread to the forest because of asymptomatic nursery plants. Thus, in the near future, significant efforts will be required to address the integrated management of this disease through the application of eco-friendly technologies.

Author details

Adugnaw Mekonene Admas* and Leykun Kebede
Ethiopian Forestry Development, Addis Ababa, Ethiopia

*Address all correspondence to: adu.biot@gmail.com

References

[1] De Rigo D, Bosco C, Durrant TH, et al. Forest resources in Europe: An integrated perspective on ecosystem services, disturbances and threats. The European Atlas of Forest Tree Species. Luxembourg: Office of the European Union; 2016. pp. 8-19

[2] Richardson DM, Rundel PW. Ecology and biogeography of Pinus: An introduction. In: Richardson DM, editor. Ecology and Biogeography of Pinus. Cambridge, UK: Cambridge University Press; 2000. pp. 3-46

[3] Holmes TP, Aukema JE, Von Holle B, et al. Economic impacts of invasive species in forests past, present, and future. In: Ostfeld RS, Schlesinger WH, editors. The Year in Ecology and Conservation Biology 2009. Annals of the New York Academy of Sciences. Vol. 1162. Wiley-Blackwell; 2009. pp. 18-38

[4] Vettraino A, Roques A, Yart A, Fan J, Sun J, Vannini A. Sentinel trees as a tool to forecast invasions of alien plant pathogens. PLoS One. 2015;**10**:e0120571

[5] Wingfield MJ, Brockerhoff EG, Wingfield BD, Slippers B. Planted forest health: The need for a global strategy. Science. 2015;**349**:832-836

[6] Liebhold AM, Brockerhoff EG, Garrett LJ, Parke JL, Britton KO. Live plant imports: The major pathway for forest insect and pathogen invasions of the US. Frontiers in Ecology and the Environment. 2012;**10**:135-143

[7] Eschen R, Britton K, Brockerhoff E, et al. International variation in phytosanitary legislation and regulations governing importation of plants for planting. Environmental Science and Policy. 2015;**51**:228-237

[8] Boyd IL, Freer-Smith PH, Gilligan CA, Godfray HCJ. The consequence of tree pests and diseases for ecosystem services. Science. 2013;**342**:1235773

[9] Klapwijk MJ, Hopkins AJM, Eriksson L, et al. Reducing the risk of invasive forest pests and pathogens: Combining legislation, targeted management and public awareness. Ambio. 2016;**45**:223-234

[10] Stenlid J, Oliva J, Boberg JB, Hopkins AJM. Emerging diseases in European forest ecosystems and responses in society. Forests. 2011;**2**:486-504

[11] Kubíková J. The surface mycoflora of ash roots. Transactions of the British Mycological Society. 1963;**46**:107-114

[12] Halmschlager E, Kowalski T. The mycobiota in nonmycorrhizal roots of healthy and declining oaks. Canadian Journal of Botany. 2004;**82**:1446-1458. Forests 2021, 12, 811 17 of 18

[13] Bartnik C. Symptomy chorobowe i uszkodzenia w systemach korzeniowych zamieraj ˛acych d˛ebów oraz grzyby im towarzysz ˛ace. In: Majewski T, editor. Fitopatologia wczoraj, dzi's i jutro; Materiały z Sympozjum 75 lat Katedry Fitopatologii SGGW w Warszawie, 23-24 wrze'snia. Warsaw, Poland: Exit; 1997. pp. 3-10

[14] Babadoost M. Fusarium: Historical and continued importance. In: Askun T, editor. Fusarium–Plant Diseases, Pathogen Diversity, Genetic Diversity, Resistance and Molecular Markers. London, UK, London, UK: IntechOpen; 2018. pp. 13-24

[15] James RL, Dumroese RK. Investigations of Fusarium diseases

within inland Pacific northwest Forest nurseries. In: Jackson WY, Ogden UT, editors. uyon J.C. Comp. 2006. Proceedings of the 53 Western International Forest Disease Work Conference, Jackson, WY, USA, 26-29 August 2005. Jackson, WY, USA: U.S. Department of Agriculture, Forest Service, Intermountain Region; 2006. pp. 3-11

[16] Martin-Pinto P, Pajares JA, Pando V, Diez JJ. Fungi isolated from diseased nursery seedlings in Spain. New Forest. 2006;**31**:41-56

[17] Menkis A, Vasiljauskas R, Taylor AFS, Stenström E, Stenlid J, Finlay R. Fungi in decayed roots of conifer seedlings in forest nurseries, afforested clear-cuts and abandoned farmland. Plant Pathology. 2006;**55**:117-129

[18] Fajardo MA, Leon JD, Correa GA, Morales JG. The causal agent of damping-off Pinus patula (Schiede) and *Pinus tecunumanii* (Schwerdtf.). Floresta Ambient. 2019;**26**:e20190050

[19] Wingfield MJ, Hammerbacher RJ, Ganley RJ, Steenkamp E, Gordon TR, Wingfield BD, et al. Pitch canker caused by *Fusarium circinatum*-A growing threat to pine plantations and forests worldwide. Australasian Plant Pathology. 2008;**37**:319-334

[20] Berg G, Hallmann J. Control of plant pathogenic fungi with bacterial endophytes. In: Schulz BJE, Boyle CJC, Sieber TN, editors. Microbial Root Endophytes. Berlin: Springer; 2006. pp. 53-69. DOI: 10.1007/3-540-33526-9_4

[21] Dumroese RK, James RL. Root diseases in bareroot and container nurseries of the Pacific Northwest: Epidemiology, management, and effects on out planting performance. New Forest. 2005;**30**:185-202

[22] Chavarriaga D, Bodles WJA, Leifert C, Belbahri L, Woodward S. *Phytophthora cinnamomi* and other fine root pathogens in north temperate pine forests. FEMS Microbiology Letters. 2007;**276**:67-74

[23] Kwasna H, Behnke-Borowczyk J, Gornowicz R, Łakomy P. Effects of preparation of clear-cut forest sites on the soil mycobiota with consequences for scots pine growth and health. Forest Pathology. 2019;**49**:e12494

[24] Newsham KK. Response of saprotrophic fungal communities to declining SO_2 pollution in the natural environment. Pedobiologia. 2003;**47**:77-84

[25] Yamazaki M, Iwamoto S, Seiwa K. Distance- and density-dependent seedling mortality caused by several diseases in eight ng in a temperate forest. Plant Ecology. 2009;**201**:181-196

[26] Shi L, Dossa GGO, Paudel E, Zang H, Xu J, Harrison RD. Changes in fungal communities across a forest disturbance gradient. Applied and Environmental Microbiology. 2019;**85**:e00080-e00019

[27] Axelrood PE, Chapman WK, Seifert KA, Trotter DB, Shrimpton G. Cylindrocarpon and Fusarium root colonization of Douglas-fir seedlings from British Columbia reforestation sites. Canadian Journal of Forest Research. 1998;**28**:1198-1206

[28] Chehri K, Salleh B, Soleimani MJ, Reddy KRN, Zakaria L. Occurrence of Fusarium spp. associated with root tissues and rhizosphere soils of forest trees and assessment of their pathogenicity on *Prunus amygdalus* seedlings. Australian Journal of Botany. 2010;**58**:679-686

[29] Mwanza EJM, Kellas JD. Identification of the fungi associated

with damping-off in the regeneration of *Eucalyptus obliqua* and *E. radiata* in a central Victorian forest. European Journal of Forest Pathology. 1987;**17**:237-245

[30] Mazarotto EJ, Poitevin CG, do Carmo ALM, dos Santos AF, Tralamazza SM, Pimentel IC. Pathogenic Fusarium species complexes associated to seeds of indigenous Brazilian forest tree *Aspidosperma polyneuron*. European Journal of Plant Pathology. 2020;**158**:849-857

[31] Weidensaul TC, Wood FA. Sources of species of Fusarium in Northern hardwood forests. Phytopathology. 1973;**63**:367-371

[32] Summerell BA, Rugg CA, Burgess LW. Mycogeography of Fusarium: Survey of Fusarium species associated with forest and woodland communities in North Queensland, Australia. Mycological Research. 1993;**97**:1015-1019

[33] Kannangara BTSDP, Deshappriya N. Microfungi associated with leaf litter decomposition of *Michelia nilagirica* and Semecar-puscoriacea at Hakgala Montane Forest. Journal of the National Science Foundation of Sri Lanka. 2005;**33**:81-91

[34] Sharma G, Pandey RR, Singh MS. Microfungi associated with surface soil and decaying leaf litter of *Quercus serrata* in a subtropical natural oak forest and managed plantation in Northeastern India. African Journal of Microbiology Research. 2011;**5**:777-787

[35] Manshor N, Rosli H, Ismail NA, Salleh B, Zakaria L. Diversity of Fusarium species from highland areas in Malaysia. Tropical Life Sciences Research. 2012;**23**:1-15

[36] Allegrucci N, Bucsinszky AM, Arturi M, Cabello MN. Comunities of anamorphic fungi on green leaves and leaf litter of native forests of *Scutia buxifolia* and *Celtis tala*: Composition, diversity, seasonability and substrate specificity. Revista Iberoamericana de Micología. 2015;**32**:71-78

[37] Rambey R, Sianturi SD. Decomposition rate of Rhizopora stylosa litter in Tanjung Rejo Village, deli Serdang Regeney, north Sumatera Province. In: International Conference of Agriculture, Environment and Food Security, IOP Conference Series: Earth and Environmental Science; IOP. Vol. 122. London, UK; 2008. p. 012058

[38] James RL. Damping-off. In: Cram MM, Frank MS, Mallams KM, tech. coords. Forest Nursery Pests. Agriculture Handbook 680. Washington, DC: USDA Forest Service; 2012a. pp. 115-116

[39] Jankowiak R, St. Epniewska H, Szwagrzyk J, Bilanski P, Gratzer G. Characterization of Cylindrocarpon-like species associated with litter in the old-growth beech forests of Central Europe. Forest Pathology. 2016;**46**:582-594

[40] Pérez Sierra A, Landeras E, León M, Berbegal M, Armengol J, García-Jiménez J. Characterization of *Fusarium circinatum* from Pinus spp. in Northern Spain. Mycological Research. 2007;**111**:832-839

[41] James RL. Damping-off. In: Cram MM, Frank MS, Mallams KM, tech. coords. Forest Nursery Pests. Agriculture Handbook. vol. 680. Washington, DC: USDA Forest Service; 2012. 115-116

[42] Baker KF. Damping-off and related diseases. In: Baker KF, editor. The U.C. System for Producing Healthy Container-Grown Plants. Calif. Agric. Exper. Sta. Ext. Serv. Manual 23. Parramatta, Australia: Australian Nurserymen's Association Ltd; 1957. pp. 34-51

[43] Peterson GW. Disease problems in the production of containerized forest tree seedlings in North America. In: Tinus RW, Stein I, Balmer WE, editors. Proceedings, North American Containerized Forest Tree Seedling Symposium. Vol. 68. Great Plains Agricultural Council Pub; 1974. pp. 170-172

[44] Dequn Z, Sutherland JR. Diseases of eucalyptus forest nursery seedlings and their management in forest nurseries in Yunnan Province, China. In: Perrin R, Sutherland JR, editors. Diseases and Insects in Forest Nurseries. Vol. 68. Paris: Institut National de la Recherche Agronomique. INRA; 1994. pp. 45-49

[45] Vaartaja O, Cram WH. Damping-off pathogens of conifers and of Caragana in Saskatchewan. Phytopathology. 1956;**46**:391-397

[46] Ferreira FA, Garcia MCC, Sanfuentes EVS. Evaluation of Fungi for Biocontrol of *Cylindrocladium Scoparium*, *Rhizoctonia Solani* and *Botrytis Cinferea* to be Used in Suspended Brazilian Eucalyptus Nurseries. 1997

[47] Singh O, Chaukiyal SP, Sharma HP. Control against damping off in spruce nurseries. Indian Journal of Forestry. 1985;**8**(4):321-322

[48] Starkey T, Enebak SA. Rhizoctonia blight of southern pines. In: Cram MM, Frank MS, Mallams KM, tech. coords. Forest Nursery Pests. Agriculture Handbook. vol. 680. Washington (DC): USDA Forest Service; 2012. 680 pp. 63-65

[49] Lilja A, Hallaksela AM, Heinonen R. Fungi colonizing scots pine cone scales and seeds and their pathogenicity. European Journal of Forest Pathology. 1995;**25**(1):38-46

[50] Mazodze R. An overview of forest nursery diseases and insects in Zimbabwe. In: Perrin R, Sutherland JR, editors. Diseases and Insects in Forest Nurseries. Vol. 68. Paris: Institut National de la Recherche Agronomique. INRA; 1994. pp. 293-297

[51] Motta E, Perrin R. Phoma herbarum and Phomopsis occulta, seed-borne pathogens causing dampingoff of larch. In: Perrin R, Sutherland JR, editors. Diseases and Insects in Forest Nurseries. Vol. 68. Paris: Institut National de la Recherche Agronomique. INRA; 1994. pp. 93-101

[52] Remadevi OK, Nagaveni HC, Muthukrishnan R. Pests and diseases of sandalwood plants in nurseries and their management. In: Lilja A, Sutherland JR, Poteri M, Mohanan C, editors. Diseases and Insects in Forest Nurseries - Proceedings of the 5th Eeting of IUFRO Working Party S7.03.04, Helsinki: Working Papers of the Finnish Forest Research Institute 11. Available from: http://www.metla.fi/julkaisut/workingpapers/2005/mwp011.htm. 2005 [Accessed: January 2, 2013]

[53] Sahashi N, Kubono T, Shoji T. Pathogenicity of *Colletotrichum dematium* isolated from current-year beech seedlings exhibiting damping-off. European Journal of Forest Pathology. 1995;**25**(3):145-151

[54] James RL. Diseases of conifer seedlings caused by seed-borne Fusarium species. In: Shearer RC, editor. Proceedings: Conifer Tree Seed in the Inland Mountain West Symposium. Ogden, UT: USDA Forest Service, Intermountain Research Station. General Technical Report GTR-INT-203:; 1986. pp. 267-271

[55] Stepniewska H. Phytophthora spp. on beech seedlings in some forest nurseries of South Poland. In: Phytophthora Spp. in Nurseries and Forest Stands. Warsaw,

Poland: Forest Research Institute; 2005. pp. 47-52

[56] Unestam T, Beyer-Ericson L, Strand M. Involvement of *Cylindrocarpon destructans* in root death of pinus sylvestris seedlings: Pathogenic behaviour and predisposing factors. Scandinavian Journal of Forest Research. 1989;**4**(4):521-535

[57] Zethner O, Jorgensen J, Husaeni EA. The current status of diseases and pests of forest tree seeds in south East Asia, especially diseases in Indonesia. In: ISTA Tree Seed Pathology Meeting, Proceedings, 1996. International Seed Testing Association; 1997. pp. 86-94

[58] Gromovykh TI, Malinovsky AL, Gukasyan BM, Tulpanova BA, Golovanova TI. The application of microbiological antagonism for biological control of damping-off in nurseries. In: James RL, editor. Proceedings of the 3rd Meeting of IUFRO Working Party S7.03.04. Missoula (MT): USDA Forest Service, Northern Region. Forest Health Protection Report 97-4; 1997. pp. 138-144

[59] James RL. Fusarium root and stem diseases. In: Cram MM, Frank MS, Mallams KM, tech. coords. Forest Nursery Pests, Agriculture Handbook. vol. 680. Washington (DC): USDA Forest Service; 2012b. pp. 117-120

[60] Martin-Pinto P, Pajares J, Diez J. Pathogenicity of *Fusarium verticillioides* and *Fusarium oxysporum* on Pinus nigra seedlings in Northwest Spain. Forest Pathology. 2008;**38**:78-82

[61] Yang D, Bernier L, Dessureault M. Phaeotheca dimorphospora increases *Trichoderma harzianum* density in soil and suppresses red pine damping-off caused by *Cylindrocladium scoparium*. Canadian Journal of Botany. 1995;**73**(5):693-700

[62] Britton KO. Damping-off of flowering dogwood seedlings caused by *Colletotrichum acutatum* and *Fusarium oxysporum*. Plant Disease. 1995;**79**(11):1188

[63] Camporota P, Perrin R. Survey for damping-off in forest nurseries in France. Preliminary results. In: Perrin R, Sutherland JR, editors. Diseases and Insects in Forest Nurseries. Vol. 68. Paris: Institut National de la Recherche Agronomique. INRA; 1994. pp. 51-58

[64] Weiland JE. Pythium root rot. In: Cram MM, Frank MS, Mallams KM, tech. coords. Forest Nursery Pests Agriculture Handbook. vol. 680. Washington (DC): USDA Forest Service; 2012a. pp. 129-131

[65] Landis TD, Tinus RW, McDonald SE, Barnett JP. The container tree nursery manual. In: Volume 5, the Biological Component: Nursery Pests and Mycorrhizae, Agriculture Handbook. Vol. 674. Washington (DC): USDA Forest Service; 1990. 171 p

[66] Fraedrich SW, Cram MM. 2012. Seed fungi. In: Cram MM, Frank MS, Mallams KM, tech. coords. Forest Nursery Pests, Agriculture Handbook. Washington (DC): USDA Forest Service. 680: 132-134

[67] Landis TD, Tinus RW, McDonald SE, Barnett JP. The container tree nursery manual. In: Volume 2, Containers and Growing Media. Agriculture Handbook. Vol. 674. Washington (DC): USDA Forest Service; 1990b. 88 p

[68] Weiland JE. Soil-pest relationships. In: Cram MM, Frank MS, Mallams KM, tech. coords. Forest Nursery Pests, Agriculture Handbook. vol. 680. Washington (DC): USDA Forest Service; 2012. pp. 16-19

[69] Anderson PK, Cunningham AA, Patel NG, Morales FJ, Epstein PR,

Daszak P. Emerging infectious diseases of plants: Pathogen pollution, climate change and agrotechnology drivers. Trends in Ecology & Evolution. 2004;**19**:535-544. DOI: 10.1016/j.tree.2004.07.021

[70] Chakraborty S. Potential impact of climate change on plant–pathogen interactions. Australasian Plant Pathology. 2005;**34**:443-448

[71] Eastburn DM, McElrone AJ, Bilgin DD. Influence of atmospheric and climatic change on plant-pathogen interactions. The Plant Pathology. 2011;**60**:54-59

[72] Colhoun J. Effect of environmental factors on plant disease. Annual Review of Phytopathology. 1973;**11**:343-364

[73] Huber L, Gillespie TJ. Modelling leaf wetness in relation to plant disease epidemiology. Annual Review of Phytopathology. 1992;**30**:553-577

[74] Deutsch CA, Tewksbury JJ, Huey RB, Sheldon KS, Ghalambor CK, Haak DC, et al. Impacts of climate warming on terrestrial ectotherms across latitude. Proceedings of the National Academy of Sciences of the United States of America. 2008;**105**(18):6668-6672. DOI: 10.1073/pnas.0709472105. Epub 2008 May 5

[75] Salinari F, Giosue S, Tubiello FN, Rettori A, Rossi V, Spanna F, et al. Downy mildew (*Plasmopara viticola*) epidemics on grapevine under climate change. Global Change Biology. 2006;**12**:1299-1307

[76] Watt MS, Palmer DJ, Bulman LS. Predicting the severity of Dothistroma needle blight on Pinus radiata under future climate in New Zealand. New Zealand Journal of Forestry Science. 2011;**41**:207-215

[77] Chakraborty S, Datta S. How will plant pathogens adapt to host plant resistance at elevated CO_2 under a changing climate? New Phytologist. 2003;**159**:733-742

[78] Garbelotto M, Linzer R, Nicolotti G, Gonthier P. Comparing the influences of ecological and evolutionary factors on the successful invasion of a fungal forest pathogen. Biological Invasions. 2010;**12**:943-945

[79] Davis MB, Shaw RG. Range shifts and adaptive responses to quaternary climate change. Science. 2001;**292**:673-679

[80] Hannukkala AO, Kaukoranta T, Lehtinen A, Rahkonen A. Late-blight epidemics on potato in Finland, 1933-2002; increased and earlier occurrence of epidemics associated with climate change and lack of rotation. Plant Pathology. 2007;**56**:167-176

[81] Bentz BJ, Régnière J, Fettig CJ, Hansen EM, Hayes JL, Hicke JA, et al. Climate change and bark beetles of the Western United States and Canada: Direct and indirect effects. Bioscience. 2010;**60**:602-613. DOI: 10.1525/bio.2010.60.8.6

[82] Woods AJ, Heppner D, Kope HH, Burleigh J, Maclauchlan L. Forest health and climate change: A British Columbia perspective. The Forestry Chronicle. 2010;**86**:412-422

[83] Fitt BDL, Fraaije BA, Chandramohan P, Shaw MW. Impacts of changing air composition on severity of arable crop disease epidemics. Plant Pathology. 2011;**60**:44-53. DOI: 10.1111/j.1365-3059.2010.02413.x

[84] Siebold M, von Tiedemann A. Potential effects of global warming on oil seed rape pathogens in Northern

Germany. Fungal Ecology. 2012;**5**:62-72. DOI: 10.1016/j.funeco.2011.04.003

[85] Ghini R, Hamada E, Bettiol W. Climate change and plant diseases. Scientia Agricola. 2008;**65**: 98-107. DOI: 10.1590/S0103-90162008000700015

[86] Compant S, van der Heijden MGA, Sessitsch A. Climate change effects in beneficial plant-microorganism interactions. FEMS Microbiology Ecology. 2010;**73**:197-214. DOI: 10.1111/j.1574-6941.2010.00900.x

[87] Siddiqui ZA, Mahmood I. Effects of rhizobacteria and root symbionts on the reproduction of *Meloidogyne javanica* and growth of chickpea. Bioresource Technology. 2001;**79**:41-46

[88] Downing KJ, Thomson JA. Introduction of the *Serratia marcescens* chi A gene into an endophytic *Pseudomonas fluorescens* for the biocontrol of phytopathogenic fungi. Canadian Journal of Microbiology. 2000;**46**:363-369

[89] Sturz AV, Christie BR, Nowak J. Bacterial endophytes: Potential role in developing sustainable systems of crop production. Critical Reviews in Plant Sciences. 2000;**19**:1-30

[90] Adhikari TB, Joseph CM, Yang G, Phillips DA, Nelson LM. Evaluation of bacteria isolated from rice for plant growth promotion and biological control of seedling disease of rice. Canadian Journal of Microbiology. 2001;**47**:916-924

[91] Tjamos EC, Tsitsigiannis DI, Tjamos SE, Antoniou P, Katinakis P. Selection and screening of endorhizosphere bacteria from solarised soils as biocontrol agents against *Verticillium dahliae* of solanaceous hosts. European Journal of Plant Pathology. 2004;**110**:35-44

[92] Selim NA, Abdel Magied AH, Habib HH, Waly HA, Fadl AA, Shalash SM. Effect of pectinase enzyme supplementation and low energy corn-soybean meal diets on broiler performance and quality of carcass and meat. Egyptian Poultry Science. 2016;**36**:319-335

[93] Goudjal Y, Toumatia O, Sabaou N, BarakateM,MathieuF,ZitouniA.Endophytic actinomycetes from spontaneous plants of Algerian Sahara: Indole-3-acetic acid production and tomato plants growth promoting activity. World Journal of Microbiology and Biotechnology. 2013;**29**(10):1821-1829

[94] Mathiyazhagan S, Kavitha K, Nakkeeran S, Chandrasekar G, Manian K, Renukadevi P, et al. PGPR mediated management of stem blight of *Phyllanthus amarus* (Schum and Thonn) caused by *Corynespora cassiicola* (Berk and Curt) Wei. Archives of Phytopathology and Plant Protection. 2004;**37**:183-199. DOI: 10.1080/03235400410001730658

[95] Brasier C, Webber J. Sudden larch death. Nature. 2010;**466**:824-825

[96] Wingfield JC. Comparative endocrinology, environment and global change. General and Comparative Endocrinology. 2008;**157**:207-216

[97] Gilbert GS, Parker IM. Rapid evolution in a plant pathogen interaction and the consequences for introduced host species. Evolutionary Applications. 2010;**3**:144-156

[98] Martin-Garcia J, Zas R, Solla A, Woodward S, Hantula J, Vainio EJ, et al. Environmentally friendly methods for controlling pine pitch canker. Plant Pathology. 2019;**68**:843-860

[99] Schulz AN, Lucardi RD, Marsico TD. Successful invasions and failed biocontrol: The role of antagonistic species interactions. Bioscience. 2019;**69**:711-724

[100] Halecker S, Wennrich JP, Rodrigo S, Andree N, Rabsch L, Baschien C, et al. Fungal endophytes for biocontrol of ash dieback: The antagonistic potential of *Hypoxylon rubiginosum*. Fungal Ecology. 2020;**45**:100918

[101] Zampieri E, Giordano L, Lione G, Vizzini A, Sillo F, Balestrini R, et al. A nonnative and a native fungal plant pathogen similarly stimulate ectomycorrhizal development but are perceived differently by a fungal symbiont. The New Phytologist. 2017;**213**:1836-1849

[102] Jones O, Scheuerlein A, Salguero-Gómez R, et al. Diversity of ageing across the tree of life. Nature. 2014;**505**:169-173. DOI: 10.1038/nature12789

[103] Aoki T, Smith JA, Mount LL, Geiser DM, O'Donnell K. *Fusarium torreyae* Sp. Nov., a pathogen causing canker disease of *Florida torreya* (*Tarreya taxifolia*), a critically endangered conifer restricted to Northern Florida and Southwestern Georgia. Mycologia. 2013;**105**(2):312-319. Available from: http://www.jstor.org/stable/43924156